Multi-Processor System-on-Chip 2

To my parents, sisters and husband, the loves and pillars of my life.

Liliana ANDRADE

I express my profound gratitude to Karine, my parents and all my family, for their help and support throughout all these years.

Frédéric ROUSSEAU

SCIENCES

Electronics Engineering, Field Director – Francis Balestra

Design Methodologies and Architecture,
Subject Head – Ahmed Jerraya

Multi-Processor System-on-Chip 2

Applications

Coordinated by
Liliana Andrade
Frédéric Rousseau

ISTE

WILEY

First published 2020 in Great Britain and the United States by ISTE Ltd and John Wiley & Sons, Inc.

ISTE Ltd
27-37 St George's Road
London SW19 4EU
UK

www.iste.co.uk

John Wiley & Sons, Inc.
111 River Street
Hoboken, NJ 07030
USA

www.wiley.com

Library of Congress Control Number: 2020940076

British Library Cataloguing-in-Publication Data
A CIP record for this book is available from the British Library
ISBN 978-1-78945-022-4

ERC code:
PE6 Computer Science and Informatics
PE6_1 Computer architecture, pervasive computing, ubiquitous computing
PE6_10 Web and information systems, database systems, information retrieval and digital libraries, data fusion
PE7 Systems and Communication Engineering
PE7_2 Electrical engineering: power components and/or systems

Contents

Foreword

Ahmed JERRAYA

Cyber Physical Systems Programs, CEATech, Grenoble, France

Multi-core and multi-processor SoC (MPSoC) concepts started in the late 1990s, mainly to mitigate the complexity of application-specific integrated circuits (ASICs) and to bring some flexibility. The integration of instruction-set processors into ASIC design aimed both to structure the architecture and to allow for programmability. The concept was adopted for general-purpose CPU and GPU in the second phase. Among the pioneers of MPSoC design, we can list the MPA architecture from ST that used eight specific cores to implement MPEG4 in 1998. This evolved 10 years later to give rise to MPPA, the Kalray's general-purpose MPSoC architecture. Another pioneer is the emotion engine from Sony that used five cores (two DSP and three RISC) to build the application processor for the PlayStation (PS2). This also evolved and later converged to bring the CELL architecture (developed jointly by Sony, IBM and Toshiba) in 2005. In 2000, Lucent announced Daytona (quad SPARC V8), and in 2001, Philips designed the famous Viper architecture that combined a MIPS architecture and a DSP (Trimedia). In 2004, TI introduced the OMAP architecture that combined an ARM and a DSP. Using MPSoC to build specific architectures is continuing, and almost every SoC produced today is a multi (or many) core architecture. An important evolution took place in 2005 with the ARM MPCore, the first general-purpose quad core. This was followed by several commercial, general-purpose multi-cores, including Intel Core Duo Pentium, AMD Opteron, Niagra Spark, the Cell processor (8 Cell cores + PowerPC, ring network).

MPSoC started a new computing era, but brought a twofold challenge: building multi-core HW that can be used easily by SW designers, and building distributed SW that fully exploits HW capabilities. To deal with these challenges, the design communities from Academia and Industry began a series of conferences and workshops to rethink classical distributed computing. The study of new methods, models and tools to deal with these new distributed HW and SW architectures

Multi-Processor System-on-Chip 2 – Applications,
coordinated by Liliana ANDRADE and Frédéric ROUSSEAU. © ISTE Ltd 2020.

generated new concepts, such as the interconnect architectures called network-on-chip (NoC). The MPSoC Forum, created in 2001, was the first interdisciplinary forum that brought together the leading thinkers from the different fields to design multi-core and multi-processor SoC. Over the last 20 years, MPSoC was a unique opportunity for me to meet so many of the world's top researchers and to communicate with them in person, in addition to enjoying the high-quality conference programs. The confluence of academic and industrial perspectives, and hardware and software, makes MPSoC not "yet another conference". I have learned how emerging SW and HW design technologies and architectures can benefit from advanced semiconductor manufacturing technologies to build energy-efficient multi-core architectures that can serve advanced computing (image, vision and cloud) and distributed networked systems. This book, in two volumes (Architectures and Applications), was published to celebrate the 20th anniversary of MPSoC with outstanding contributions from previous MPSoC events.

This first volume on architectures covers the key components of MPSoC: processors, memory, interconnect and interfaces.

Acknowledgments

Liliana ANDRADE and Frédéric ROUSSEAU

Université Grenoble Alpes, CNRS, Grenoble INP, TIMA, 38000 Grenoble, France

The editors are indebted to the MPSoC community who made this book possible. First of all, they acknowledge the societies that supported this project. EDAA and IEEE/CAS partially funded the organization of the first two events. Since its creation, IEEE/CEDA has sponsored the event. Industrial sponsors played a vital role in keeping MPSoC alive for the last 20 years; special thanks to Synopsys, Arteris, ARM, XILINX and Socionext. The event was created by a nucleus of several people who now form the steering committee (Ahmed Jerraya, Hannu Tenhunen, Marilyn Wolf, Masaharu Imai and Hiroto Yasuura). A larger group has, for the last 20 years, been working to form the community (Nicolas Ventroux, Jishen Zhao, Tsuyoshi Isshiki, Frédéric Rousseau, Anca Molnos, Gabriela Nicolescu, Hiroyuki Tomiyama, Masaaki Kondo, Hiroki Matsutani, Tohru Ishihara, Pierre-Emmanuel Gaillardon, Yoshinori Takeuchi, Tom Becnel, Frédéric Pétrot, Yuan Xie, Koji Inoue, Masaaki Kondo, Hideki Takase and Raphaël David). The editors would like to acknowledge the outstanding contribution of the MPSoC speakers, and especially those who contributed to the chapters of this book. Finally, the editors would like to thank the people who participated in the careful reading of this book (Breytner Fernandez and Bruno Ferres).

PART 1

MPSoC for Telecom

1

From Challenges to Hardware Requirements for Wireless Communications Reaching 6G

Stefan A. DAMJANCEVIC[1], Emil MATUS[1], Dmitry UTYANSKY[2], Pieter VAN DER WOLF[3] and Gerhard P. FETTWEIS[1,4]

[1] *Vodafone Chair Mobile, Communications Systems, TU Dresden, Germany*
[2] *Synopsys, Saint Petersburg, Russia*
[3] *Solutions Group, Synopsys, Inc., Eindhoven, The Netherlands*
[4] *Barkhausen Institut, TU Dresden, Germany*

Over the past few decades, we have seen rapid innovation in wireless communications. In particular, the IEEE 802.11 and 3GPP standardization organizations have driven data rates into the Gb/s range, enabling modern life, at home, at work and on the road. Societies of today have become dependent on this important infrastructure. The basis of this development is an infrastructure based on electronic circuits, which are driven, at heart, by very advanced *multi-processor system-on-chip* engines.

Firstly, we want to deliver a vision on what is to be expected from 6G, the next innovation wave of cellular technology. Cellular 1G was about delivering analogue voices, and digital 2G fixed it. The intention of 3G was to deliver data, but only 4G made it proficiently available for the service requirements. With 5G, we see the advent of the Tactile Internet, i.e. connecting remote controls not as point-to-point solutions but via a network. Will 6G be just a "fix" of issues left unsolved? We believe 6G

For a color version of all figures in this book, see www.iste.co.uk/andrade/multi2.zip.

Multi-Processor System-on-Chip 2 – Applications,
coordinated by Liliana ANDRADE and Frédéric ROUSSEAU.

will deliver truly more than this, and will also require many more sophisticated signal processing tasks.

Secondly, we want to analyze the computational tasks of 5G and beyond baseband processing, as well as their specification requirements. It becomes clear that the heterogeneity of computation cannot be mapped efficiently onto a homogeneous processor array. Instead, we need to find the right architecture for the right task. The reader is introduced to an extremely varied set of requirements as the services dramatically differ in terms of latency, data rate and reliability.

Thirdly, we want to give perspective and a sense of scale required from *hardware* for the previously determined corner workloads. We do this by implementing an example beyond 5G algorithm *generalized frequency division multiplexing*, with regard to those workloads, on a prototype *software* programmable *single-instruction, multiple-data* wide vector processing machine. Workloads alone are not sufficient to deduce adequate requirements for *hardware* (HW). To bridge the gap between workloads and HW requirements, we need to know how the 6G candidate waveform modulation translates workloads into HW requirements.

Conclusively, the performance–cost analysis shows a need for high flexibility. The low-end use case of the implemented algorithm easily fits within one $1\,GHz$ 512 bit vector processor core requiring $1.01 - 5.39\,MHz$ of the processor clock budget, therefore enabling seamless time multiplexing with other *software* (SW) kernels running on the core. The *carrier aggregation* (CA) high-end use case requires $921 - 5{,}505\,MHz$, which allows the prior clock-efficient version to fit on one $1\,GHz$ 512 bit vector processor core. This sets the stage for the system designer to opt between a generalized vector processor or some more specialized HW accelerator engine, such as a dedicated (application-specific) HW accelerator or an *application-specific instruction set processor* (ASIP). Finally, the implemented algorithm running the *multiple-input, multiple-output* (MIMO) CA high-end use case would require a budget of $7.37 - 44.04\,GHz$, the theoretical equivalent of eight or forty-five 512 bit vector processor cores, making the high-end use case more suitable for execution on the dedicated HW accelerator or the ASIP, which were otherwise powered down. The use case corners demonstrate a high variability requirement from HW, which makes heterogeneous *multi-processor system-on-chip* (MPSoC) solutions ideal future-proof HW for beyond 5G. In this chapter, we present our key findings that connect the dots from vision to future HW in wireless communications.

1.1. Introduction

As we are writing this chapter in early 2020, it is obvious that there exists a gap between the conventional 5G vision (Fettweis 2012; NGMN Alliance 2015; Qualcomm 2016) and the deployed 5G. We do not have coordinated *unmanned areal vehicles* (UAV) groups humming over our cities like busy worker bees around hives (ARIB *et al.* 2016). We do not have critical *vehicle to everything* (V2X) communication with 1 ms end-to-end latency coordinated "emergency trajectory adjustment" (3GPP 2018d) keeping us safe, nor the smart infrastructure that would make traffic lights obsolete (Fettweis 2012). We do not have the cellular *virtual reality* (VR) and *augmented reality* (AR) (NGMN Alliance 2015) helping us to acquire and select important information about our environment when we explore new places. Let us investigate why is that so. How did we come into this situation, where are we now and what should we do to traverse this rift?

1G was a great step, which created the vision of ubiquitous voice telephony, but we needed 2G to deliver on the expectations created such as, for example, national and international roaming. The 3G standards were a great step towards ubiquitous cellular data, but we needed 4G to fix the problems. Now 5G should be an infliction point in bringing cellular data to new applications. However, do we need to use the 5G system to understand what is really needed, and have to wait for 6G to fix the issues? And are these fixes required to make the Tactile Internet a reality?

As we see 5G unfold, expectations on the economic and societal impact are very high. Many new opportunities for business will emerge with the new communications epoch. Besides Gb/s data rates, the Tactile Internet is the most highlighted promise of 5G, enabling remote control applications over cellular data. We will review opportunities and their technical requirements. This helps us to build an understanding, to detect missing pieces. It is of particular interest to see the broad set of chances in semiconductors unfolding before us, making the pathway to 6G maybe the next wild and open opportunity for entrepreneurship and change in company economics.

As of the first half of the 2010s, different sectors of the industry, research and, later, politicians, as well as regulatory bodies, overwhelmingly adopted and unanimously agreed that 5G was needed. The basic driver for creating the Tactile Internet vision for 5G was published in early 2014 (Fettweis 2014) and later acknowledged in (ARIB *et al.* 2016; 3GPP 2018d). The overall consensus on the need for 5G resulted in the typical standardization iterations of the 5G standard in 3GPP (2017, 2018c), landing firmly on the 5G stand-alone (3GPP 2019e, c, d). The main innovation, next to higher data rates, is the introduction of *ultra-reliable low latency communications* (URLLC) around the idea of the Tactile Internet, which again addresses a new domain not served by cellular data so far. As we slowly start understanding the true requirements and impact of URLLC, it turns out that 5G will not fully deliver a solution as required.

Given that 3GPP 5G NR specifications (3GPP 2019c, d) regarding handsets for *frequency range 1* (FR1) ($0.45\ GHz - 6\ GHz$) and *frequency range 2* (FR2) ($24.25\ GHz - 52.6\ GHz$) operating ranges exist, the MPSoCs, which can be scaled in data rate as well as latency, are still to be designed.

For the contemporary reader, this chapter offers an incremental contribution to the fulfillment of that vision, spanning the gap between 5G and 6G. For the future reader, this chapter offers a methodology of translating workloads via a specific algorithm to HW.

Section 1.2 shows the trends and analyzes the workloads defined by the standards. In addition, it gives lower and upper requirements to be considered when implementing a communications modem. In section 1.3, we give the reader a background on the 6G candidate waveform modulation *generalised frequency division multiplexing* (GFDM), develop a GFDM-associated dataflow processing graph and pseudo-code. Section 1.4 covers precision requirements, in which we explore required bit-lengths to represent data and satisfy 3GPP LTE/NR error requirements. Section 1.5 presents the implementation, GFDM vectorization variants, loop order variants and many properties that arise from loop and vectorization arrangements, such as the possibility of minimizing cycle counts for maximum throughput or minimizing the number of memory accesses for low-power operation.

1.2. Breadth of workloads

The first step towards estimating the HW requirements for a typical beyond *3GPP 5th Generation New Radio* (5G) algorithm is sizing the span of workloads in terms of throughput and deadlines under which data processing has to finish. To study the matter, we go over the holistic vision and trends of what we expect in 6G, followed by an analysis of the key deployed 5G standard specifications. From the latter, we identify the far corners that stretch the workload requirement space, see how these corners fit into the vision and trends and provide numerical values for HW requirements of a future MPSoC. Finally, throughout the workload analysis, the emerging theme is the required high flexibility, which leads us to pick a *vector digital signal processor* (vDSP) as a middle ground that provides both flexibility and good data throughput for *single instruction, multiple data* (SIMD) processing.

1.2.1. *Vision, trends and applications*

The trend in cellular communications over the past generations was to pioneer with a technology in one generation and to optimize it in the next. As mentioned in the abstract, 1G and 2G introduced and optimized voice, 3G and 4G introduced and optimized broadband data streaming and, lastly, 5G and 6G introduce and should optimize the Tactile Internet. Furthermore, the generations also had their killer applications that captured the mass market: for 3G, it was a video call; for 4G, it was social media and streaming; for 5G and 6G, we cannot say for certain, but we can make informed predictions.

As of the advent of 4G, cellular networks are not built for low latency, and the introduction of the low-latency connection between network end points as a key requirement is a challenge for cellular network operators. With the current network infrastructure, it is hard to provide both massive data rates and low latency. Early solutions revolve around trading off one for the other. Hence, in the early adaptation stages, we can expect a killer application that requires low end-to-end latency, but does not require high data rates or vice versa. Likely candidates for *low rate - low latency* (LRLL) are applications such as factory automation, remote control, trajectory alignment and emergency stop aspects of self-driving for cars, UAVs and robots across a broad field of industries, from construction sites to warehouses. On the other hand, for *high rate - high latency* (HRHL), a killer application would likely be 8K streaming or other massive data transfers, where latency is not a constraint. Drawn from experience, we can expect 6G to be able to fix and provide both low latency and massive data rates. A likely candidate to be the killer application for *high rate - low latency* (HRLL) is seamless cellular AR, VR and other Internet gaming-oriented applications. Of course, we can also expect new LRLL and HRHL applications to emerge and further develop in their respective niches. We borrow from the work of (Fettweis *et al.* 2019) the neatly organized overview and categorization of applications in Figure 1.1.

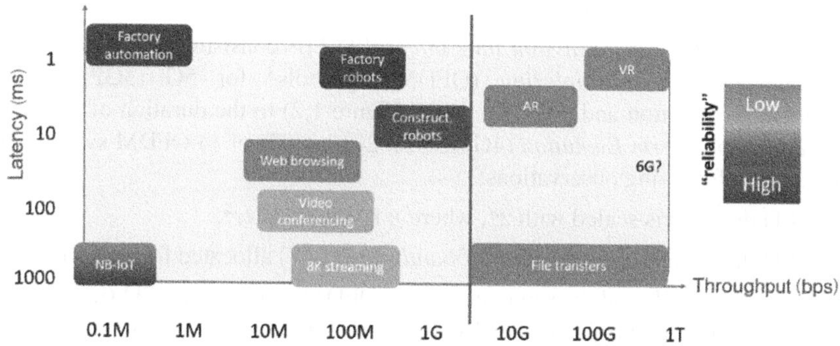

Figure 1.1. *Application mapping on the rate–latency plane with regard to the reliability requirement (Fettweis* et al. *2019)*

The reliability requirement, as part of the Tactile Internet, affects HW through new procedures and algorithms, which ultimately lead to expanded workloads in terms of additional data throughput and shorter deadlines for processing that data. Potentially, even new operating systems for MPSoCs centered on threat insulation and security will need to be investigated, which is outside the scope of this chapter, but the conclusions are the same. The prospective future 6G applications span vastly different data rates (10 kb/s – 1 Tb/s, i.e. $10^8\times$ span) and latency requirements (2 ms – 2 s i.e. $10^3\times$ span).

We have a broad and dynamic workload space with the total number of applications steadily increasing. These applications are sometimes concurrent in their operation, and in, at other times, they are exclusive, which adds another layer of complexity into consideration. For example, a handset user is either watching a video stream or playing an AR-based game while taking a stroll through the city, not both at the same time.

The challenges presented above, combined with new signal processing tasks associated with radio issues when working in new frequency ranges, make HW design choices particularly hard, efficiently balancing between architecture complexity and size, power consumption and cost. The HW challenge is certainly the "half a trillion dollar question"[1]. There is no other way to treat this than as an onion problem, and to peel it layer by layer, going step by step, the first step being standard specifications.

1.2.2. *Standard specifications*

High variability is built into modern standards through their many modes of operation. One aspect by which these modes may differ are workloads. Let us analyze the workloads by looking at the most advanced 5G standard.

1.2.2.1. *Processing deadline variability*

If we look at the *transmission time interval* (TTI)[2] consisting of *14 Orthogonal Frequency Division Multiplexing* (OFDM) symbols[3] for 5G (3GPP 2019e), extrapolate its duration and compare it (see Figure 1.2) to the duration of *3GPP 4th Generation Long Term Evolution* (4G) (3GPP 2018a) TTI of 14 OFDM symbols, we can make the following observations:

1) TTI duration is scaled with 2^{μ}, where μ is a parameter[4];

2) TTI duration is not a function of *bandwidth* (BW) allocated for that TTI;

3) for a fixed BW and the same number of OFDM symbols per TTI (i.e. the same amount of data), the duration of the TTI differs (i.e. the deadline to process that data differs);

4) there is a 16× difference in processing deadlines between the corner cases.

Depending on the mode of operation, the computational engine may need to process the same amount of data with deadlines shifting 16× during operation.

1 According to data (GFK 2019), smartphone and related wearables were a $522 billion dollar market in 2018.

2 Loosely translated as a single packet of data, for the broader audience.

3 Simplified: TTI consists of OFDM symbols, which, in turn, consist of transformed sets of *quadrature amplitude modulation* (QAM) symbols, which, in turn, consist of transformed sets of soft bits, which are a mix of guard and information bits.

4 TTI is inversely scaled with *subcarrier frequency spacing* (SCS). In 5G, SCS is base spacing ($15\,kHz$, same as 4G) $\times 2^{\mu}$ and TTI is base TTI ($1\,ms$, same as 4G) $\times 2^{-\mu}$.

The processing deadline for baseband physical layer processing is constrained by the *hybrid automatic repeat request* (HARQ) *media access control layer* (MAC-L) procedure, which is 3 TTIs long. Let us assume, as a rule of thumb, that we have $1/3$ of the 3 TTI budget associated with waveform modulation, while the other $2/3$ are reserved for other processing steps[5]. With that in mind, the deadlines are simplified and match the TTI duration for the given subcarrier spacing parameter μ.

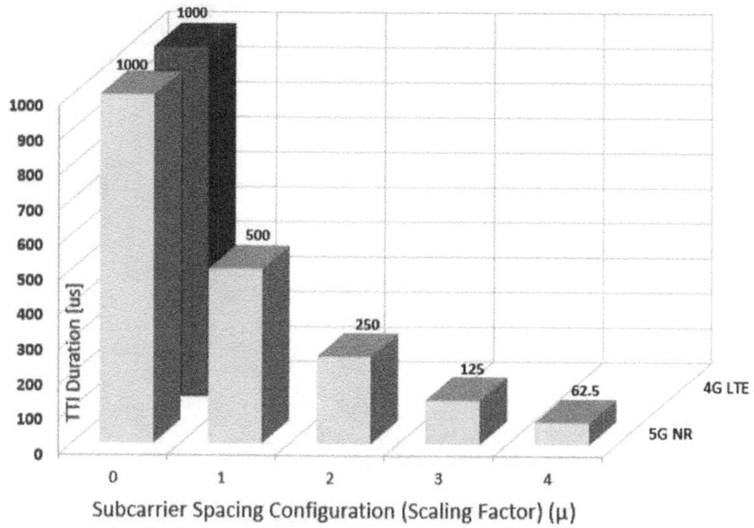

Figure 1.2. *Comparing 14 OFDM symbols' TTI duration of 4G and 5G*

1.2.2.2. *Data throughput variability*

Next, let us investigate throughput requirements of the 5G specifications. At this point, we choose to compute and show requirements per handset modem chip. Note that there are additional device classes that support only a subset of the operating modes shown here. However, an advanced handset of the future should support all of the modes shown here, to use the full potential of different frequency ranges.

Previously, we performed a specification analysis (Damjancevic *et al.* 2019), although over the past 6 months the 5G specifications have expanded, and here we show the updated information. In Figure 1.3 and Figure 1.4, we have organized the throughput information and presented it in a readable form for FR1 and FR2, respectively. For comparison, we plot also the 4G data throughput requirement, which

5 We do not have a reference for this rule of thumb, since the actual timing budget distribution is vendor-specific and kept private. This is a first-order approximation.

coexists in FR1, along with other legacy standards. Future FR2 5G systems will coexist with Super High Frequency (SHF) and Extremely High Frequency (EHF) communication and radar systems, which are region-specific, adding an extra layer of flexibility. Throughput is shown in maximum *resource blocks* (RB)[6] over time, per channel BW for one spatial MIMO data stream layer[7,8] in compliance with active (3GPP 2019c, d) specifications. From the figures, we can observe that:

1) there are many possible modes of operation;

2) there is a 352× difference between the processing data load corners (LTE, $1.4\,MHz$) – lower end and ($\mu = 3,\ 400\,MHz$) upper end in terms of RBs.

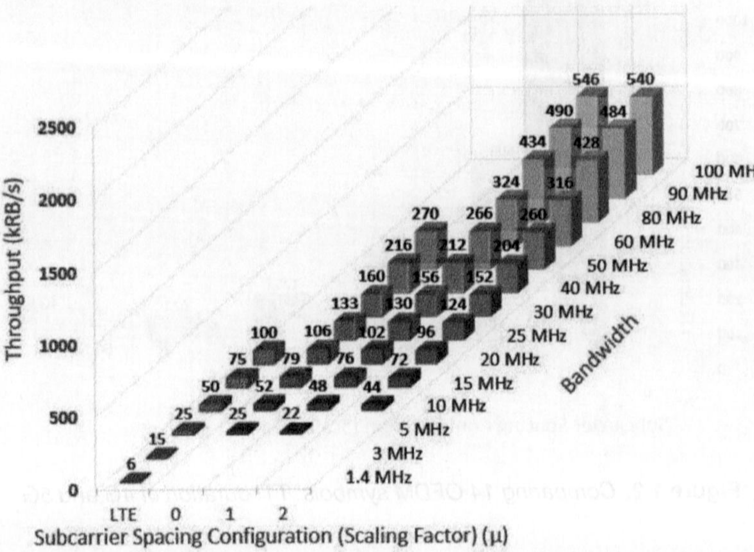

Figure 1.3. *Processing load in kRB/s for 5G NR FR1 (Damjancevic* et al. *2019)*

We see a greater need for flexibility again emerging from the observations, compared to the 4G standard, with many more modes to support on top of the throughput difference. Now that we have identified the throughput corners in RBs,

6 $1\,kRB = 1{,}000\,RB, 1\,RB = 12\,subcarriers \times 2\,slots\,of\,7\,(6)\,symbols\,per\,slot = 168\,(144)\,RE$, $(\cdot)$ – extended *Cyclic Prefix* (CP) mode possible for $\mu = 2$ or LTE. An RB is a unit in technical terminology representing a grid section onto which QAM symbols are mapped; one element of that grid is called an *resource element* (RE).

7 Simplified: for a radio link.

8 Note that this should not be confused with massive MIMO, which uses constructive and destructive interferences of radio waves from multiple antennas to create a single data stream layer.

we can assign and calculate the smallest and highest QAM and code bit rates allowed by the 4G[9] and 5G[10] specification sets, to the lower and upper ends, respectively.

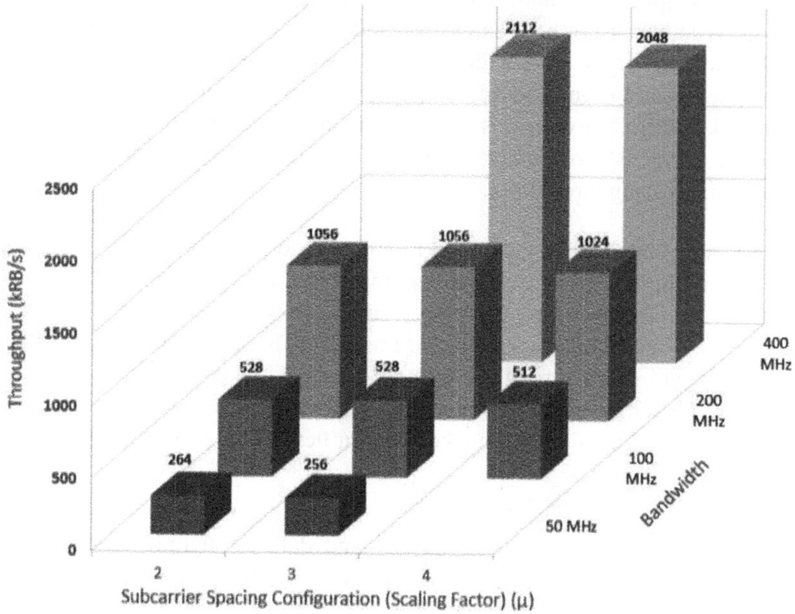

Figure 1.4. *Processing load in kRB/s for 5G NR FR2*

This sets the low end at 200 kb/s per 1.4 MHz BW channel and spatial data layer[11]. Note that this is a hard bit rate, and the rate at different processing steps may be higher due to oversampling.

The high end, on the other hand, is set at 2.63 Gb/s per 400 MHz BW channel and spatial data layer. Note that we have insisted on emphasizing the "per channel, per layer", since handsets have many operating modes and some of those modes may require a multitude of channels or layers active simultaneously, for example, CA[12]. The 5G standard allows modes that support a multitude of each, further increasing the effective number of RBs communicated. The extra RBs can be used to increase

9 4G *modulation code rate schemes* (MCS) (3GPP 2019b).

10 5G MCS (3GPP 2019f).

11 This is the lowest rate per smallest channel. Even though a voice call may require between 8 kb/s and 80 kb/s (depending on the codec (ITU 2019a, b)), the block nature of *discrete Fourier transform* (DFT) and the required sample rate used in 4G forces the modem to work with higher average rates than those required by the application.

12 Simplified: simultaneously active multiple channels on different frequencies.

the overall throughput or to increase redundancy by sending the same data on another channel frequency or layer. 5G as of now supports up to $4 \times 400\,MHz$ CA (3GPP 2019d) for its upper-end mode and MIMO up to 8 layers of data (3GPP 2019e) extension. Whether or not both extension modes can overlap within 5G is not clear, since the two are often used with opposite goals. Namely, data layer extensions conserve the spectrum and provide throughput by reusing the same spectrum on a different link, while CA extensions use excess unused spectrum to provide throughput. However, reaching 6G, we cannot dismiss the possibility that the overlap could serve as a way of increasing throughput.

With this in mind, let us make two high-end cases, first using the CA extension and the second high-end use case overlapping both the CA and multiple data layers. Calculating the first, we have $10.52\,Gb/s$ and $84.16\,Gb/s$, respectively. These rates could be used, for example, for large file transfers. The difference between the low-end and the two high-end throughput corners is approximately $5 \times 10^4\times$ or $4 \times 10^5\times$, respectively. Therefore, the system needs to deal with vastly varying data processing loads during operation, highlighting the need for flexibility of the compute engine.

1.2.2.3. *Specification summary*

We can now combine both the latency and data throughput requirement into low- and high-end corners on the 2D plane. Low-end use case is clear: LTE legacy $1.4\,MHz$. High-end, however, has two options: either the highest data of $\mu = 3$, $BW = 400\,MHz$ or the shortest deadline $\mu = 4$. We chose the former as the combined high-end due to nonlinear scaling of operations per data point in algorithms. For example, to compute the 4096 *inverse discrete Fourier transform* (IDFT) used in $\mu = 3$, $BW = 400\,MHz$ for a single OFDM symbol, it takes more operations than to compute 2×2048 IDFT for two symbols used in $\mu = 4$, $BW = 400\,MHz$, i.e. over the same duration, there are more operations in $\mu = 3$, $BW = 400\,MHz$. The corner cases are presented in Table 1.1, with expression of throughput in other units as well, for a more reader-friendly overview.

Use Case	Throughput					TTI
	$\left[\frac{RE}{symbol \cdot BW \cdot layer \cdot TTI}\right]$	$\left[\frac{RB}{BW \cdot layer \cdot TTI}\right]$	$\left[\frac{RB}{TTI}\right]$	$\left[\frac{kRB}{s}\right]$	$\left[\frac{Gbits}{s}\right]$	$[\mu s]$
Low-end LTE legacy (3GPP 2019a, b)	72	6	6	6	$0.2\,m$	1,000
CA high-end FR2 (3GPP 2019d, f) $4 \times CA, \mu = 3, 400MHz$	3,168	264	1,056	8,448	10.52	125
MIMO CA high-end FR2 (3GPP 2019d, e, f) $8 \times 8, 4 \times CA, \mu = 3{,}400MHz$	3,168	264	8,448	67,584	84.16	125

Table 1.1. *Processing requirement corners as per standard specification*

1.2.3. *Outcome of workloads*

We see that the 3GPP specifications follow the trend and vision of 5G laid out in section 1.2.1, incorporating the variability of workloads as the central paradigm.

With throughput requirements varying by several orders of magnitude, a homogeneous HW solution would be very inefficient for both high-end and low-end use cases. Rather, a heterogeneous HW architecture that is a mixture of HW accelerator engines, banks of programmable processing elements and supporting memory systems would be efficient. Accelerator engines such as dedicated (application-specific) HW accelerators and ASIPs are ideal to deal with extreme high-end use cases and easy-to-scale low-varying algorithms or processing steps, due to their speed and efficient energy per data point consumption. While banks of programmable processing elements such as vDSPs (SIMD cores with signal processing-oriented *instruction set architecture*) and generic scalar *reduced instruction set computer* (RISC) cores are ideal to deal with moderate–high to low-end use cases and processing steps that require flexibility, for example, choosing from a set which algorithm to perform, based on the device's situational parameters and environmental conditions. Such HW is well suited for dealing with highly variable loads by powering HW modules on and off based on the current load. For example, if enough compute resources are available on the vDSPs, i.e. available idle cycles, we could run the communication kernels on the vDSPs in a time-multiplexed manner and keep the HW accelerators off.

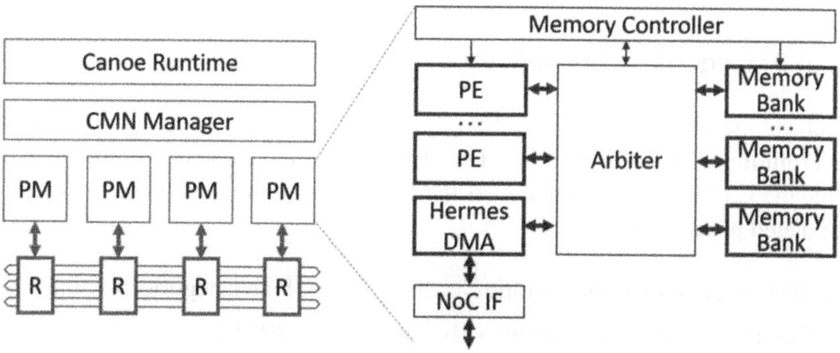

Figure 1.5. *Tiled "Kachel" MPSoC with decentralized tightly coupled memories*

When it comes to MPSoC architecture in the sense of module arrangement and layout, we have recently published works with both decentralized (e.g. in Figure 1.5, (Fettweis *et al.* 2019)) and centralized (e.g. in Figure 1.6, (Damjancevic *et al.* 2019)) memory in mind. Although this chapter follows closely the later work, the programmable vector processor is a common processing element in both, and the lessons learned are universal[13].

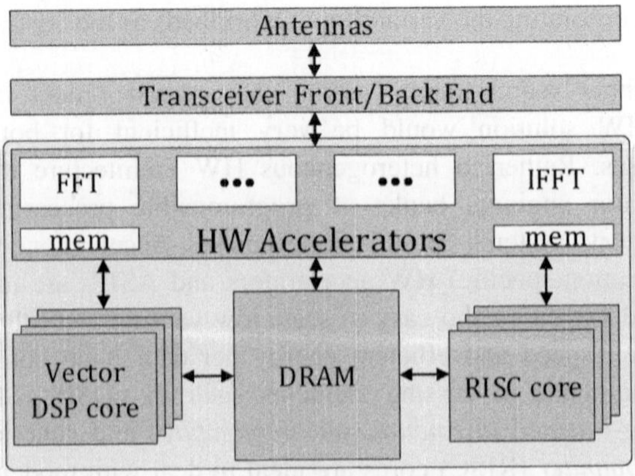

Figure 1.6. *Heterogeneous MPSoC with a central shared memory architecture*

Without discussing the layouts, as both have their advantages, let us delve into the common thread and analyze the combined effect of workloads and algorithms on HW provisioning requirements and possibly confirm our hypothesis that a heterogeneous MPSoC is required for an efficient future-proof solution.

1.3. GFDM algorithm breakdown

Knowledge of algorithm kernel[14] requirements with respect to workload is key to determining whether or not the vDSP has enough available compute resources to execute the kernel. This section covers the GFDM algorithm and its processing graph extrapolated pseudo-code.

GFDM processing in the literature is divided into three categories:

1) frequency-domain processing (Michailow *et al.* 2014);

2) time-domain processing (Farhang *et al.* 2016; Matthé *et al.* 2016);

3) Radix-2 cascade processing (Nimr *et al.* 2018).

---

13 An additional motivation for having an extent of SW solutions in communication modems is 1) their ability to be exchanged easily via updates, compared to exchanging HW, and 2) they are ideal for multiplexing a multitude of standards on the same HW.

14 A self-contained piece of SW or function that fully captures the functionality of an algorithm.

We assume (2) time-domain processing for implementation on the vDSP, primarily due to its flexibility and low compute complexity per data point on small workloads.

1.3.1. *Equation*

We consider a GFDM resource grid with K subcarriers and M symbols. In total, $N = MK$ complex data symbols are transmitted within a GFDM block, where $d_k[m]$ represents a data symbol to be transmitted on the kth subcarrier and the mth symbol for $k = 0, 1, ..., K - 1$, $m = 0, 1, ..., M - 1$. The signal corresponding to the kth subcarrier is constructed by K-times up-sampling the data vector $\boldsymbol{d}_k = [d_k[0], d_k[1], ..., d_k[M-1]]^T$, followed by pulse shape filtering with $g[mK]$ and subcarrier modulation $e^{j2\pi\frac{kn}{K}}$. The resulting GFDM signal $x[n]$ for $n = lK + i$; $l = 0, 1, \ldots, M-1$; $i = 0, 1, \ldots, K-1$ is shown in [1.1]

$$
\begin{aligned}
x[n] &= \sum_{k=0}^{K-1} \sum_{m=0}^{M-1} d_k[m]\, g[n - mK]_{mod\ N}\, e^{j2\pi\frac{kn}{K}} \\
&= \underbrace{\sum_{m=0}^{M-1} g[n - mK]_{mod\ N} \underbrace{\sum_{k=0}^{K-1} d_k[m]\, e^{j2\pi\frac{kn}{K}}}_{D_n[m] = \text{IDFT}\{d_k[m]\}}}_{GFDM\ filtering}
\end{aligned}
\qquad [1.1]
$$

where $l = 0, 1, \ldots, M-1$; $n = 0, 1, \ldots, K-1$ and $g[.]_{mod\ N}$ is the impulse response of the K-times over-sampled pulse shaping filter with circular symmetry. Reordering the sums in equation [1.1] leads to a K-point IDFT$\{\boldsymbol{d}[m]\}$ of the mth data symbol vector $\boldsymbol{d}[m] = [d_0[m], d_1[m], \ldots, d_{K-1}[m]]^T$ and subsequent filtering of IDFT outputs $\boldsymbol{D}[m] = [D_0[m], D_1[m], \ldots, D_{K-1}[m]]^T$. IDFT is a well-researched topic, hence the novelty of interest in our work is the filtering of $\boldsymbol{D}[m]$.

1.3.2. *Dataflow processing graph and matrix representation*

The associated dataflow processing graph is depicted in Figure 1.7, with the notation $foo[bar]$ abbreviated as foo_{bar}.

Each *multiply-accumulate* (MAC) in Figure 1.7 operates on input/output vector length of K elements that results in the total M^2K scalar MAC operations, $2M^2K$ loads and MK stores. This is the theoretical minimum number of operations. The MAC result is stored in an indexed accumulator register (ACC) and is used in the next MAC of the same index. For readers not familiar with dataflow diagrams, a matrix representation of GFDM filtering is shown in Figure 1.8, and the experience drawn from matrix multiplication implementations can also be used here.

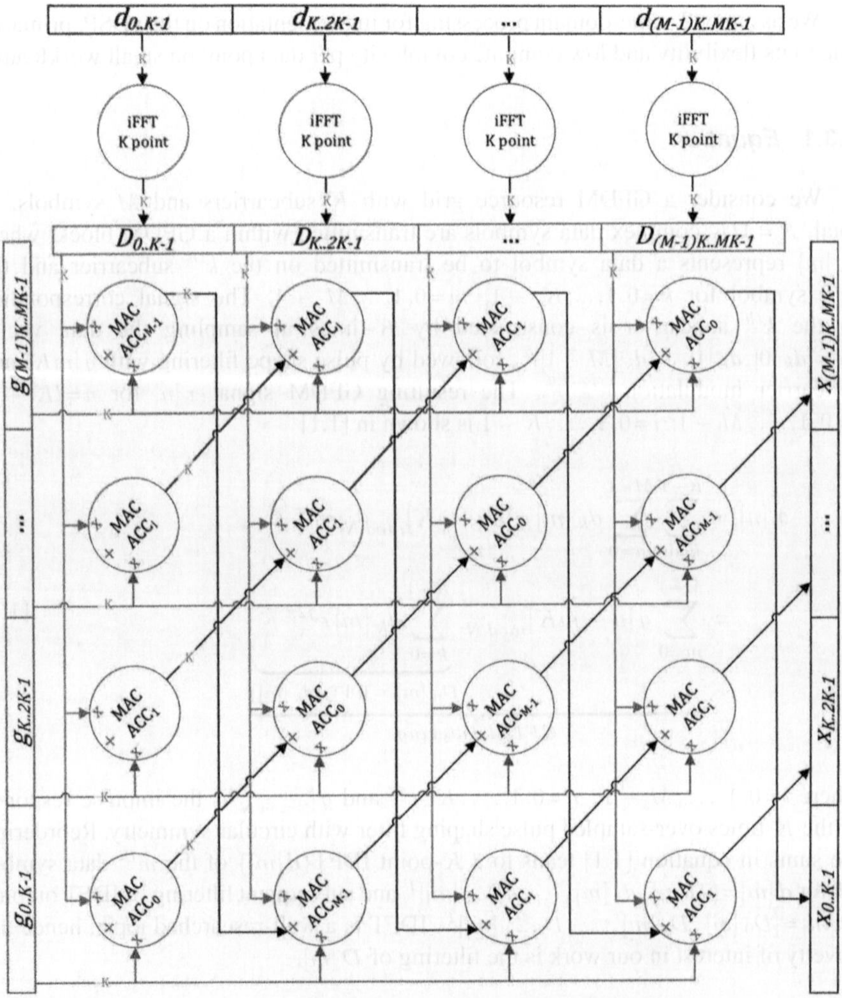

Figure 1.7. *GFDM processing dataflow diagram*

1.3.3. *Pseudo-code*

Observing only the equation, a naive programmer could see the modulo load operation on g [.] and get stuck or try to write % for modulo in their C code, hoping that the compiler could deal with it. Fortunately for us, we have drawn the processing graph and can infer the processing required and how to circumvent the modulo operation altogether.

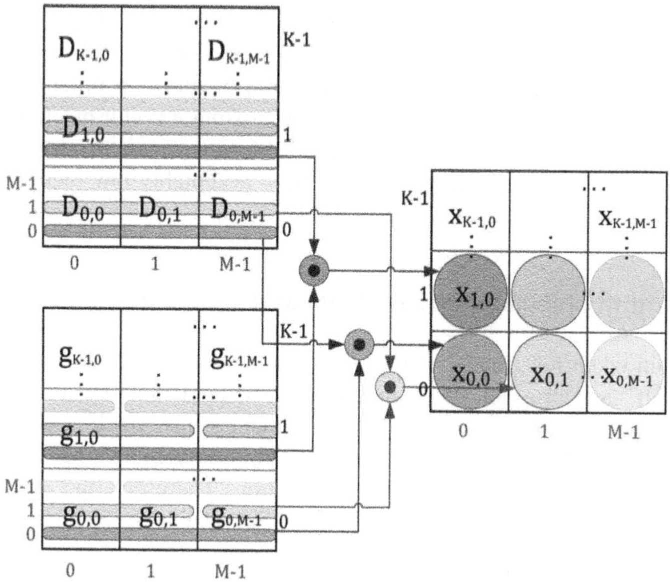

Figure 1.8. *Visualization of time-domain GFDM filtering*

```
input : IDFT output D and filter coefficients g
output: GFDM signal x
// Assume complex operations *,+=
for l ← M − 1 to 0 do
    for i ← 0 to K − 1 do
        ACC = 0;
        for m ← 0 to l do
            // split in 2 loops to avoid modulo operations
            ACC+ = D[mK + i] * g[(M − 1 + m − l)K + i];
        end
        for m ← l + 1 to M − 1 do
            ACC+ = D[mK + i] * g[(m − 1 − l)K + i];
        end
        x[(M − 1 − l)K + i] = ACC;
    end
end
```

Figure 1.9. *GFDM pseudo-code*

GFDM filtering consists of three mutually independent "for" loops (shown in Figure 1.9)[15]. Since the loops are mutually independent, they can be interchanged

15 Note that the first inner loop will not be executed for $l = 0$.

freely, which effectively changes the processing order to the programmer's desire. Again, naively, we could think that the order does not matter as long as the "job gets done", but, as we will see in section 1.5, the loop order has a notable impact on memory and processor utilization when using a vector machine. However, before we discuss different loop orders, we need to investigate fixed-point precision requirements, i.e. bit-lengths required to achieve acceptably low quantization error based on the 4G and 5G standard specification.

1.4. Algorithm precision requirements and considerations

It is often taken for granted that we can take a *state-of-the-art* (SotA) processor "off-the-shelf" and program the kernel straight without thinking much about data degradation due to quantization noise. Yet even an error as small as a percentile off the true value, compounded over all the succeeding processing steps of the system, may result in faulty data for the user application. Therefore, the standardization body defines the error budget (3GPP 2018b, 2019a), which can be distributed among different processing steps.

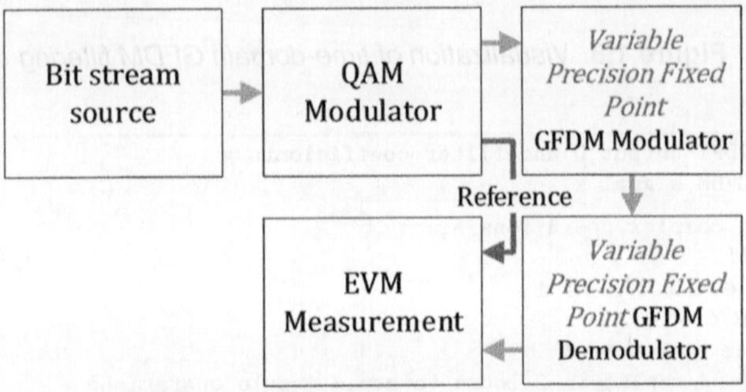

Figure 1.10. *Precision test bed set up*

Of interest is determining the necessary number of bits to represent input and output data, as well as the necessary bit-length size of the processor accumulators where intermediate results are stored. Input and output share the same data type (bit-length) since that simplifies memory access, while, for accumulators, we assume a different data type. We denote the bit-lengths for these hypothetical data types as $data_{bits}$ and ACC_{bits} for input/output and accumulators, respectively. Assuming a Q0.n: normalized to 1 signed integer notation, where n is the number of fractional bits, we consider the following test bed setup, illustrated in Figure 1.10. A random bit stream is mapped to complex QAM symbols. Next, QAM symbols are mapped to $\boldsymbol{d}_k$ in equation [1.1], a copy of $\boldsymbol{d}_k$ is kept as reference. QAM symbols are GFDM-modulated using IDFT and filtered as per the algorithm shown in Figure 1.9 and afterwards demodulated. Demodulation is done by repeating the algorithm shown in

Figure 1.9 with a new set of coefficients that invert the first filtering, followed by DFT. Finally, after successful demodulation, the demodulated stream is compared with the reference and the *error vector magnitude* is measured.

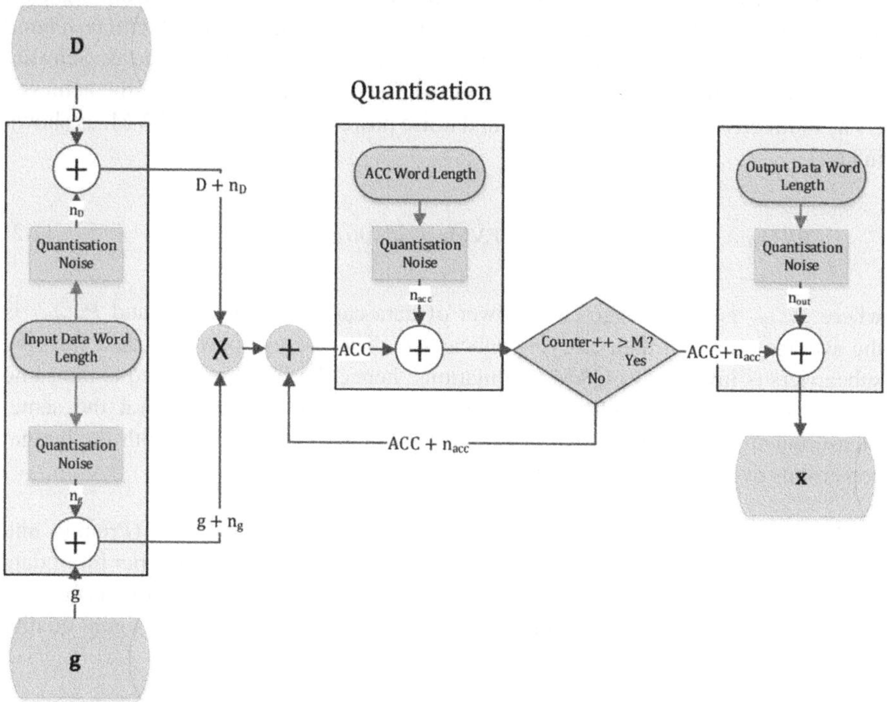

Figure 1.11. *Varied precision quantization of GFDM*

In Figure 1.11, we show the basic dataflow of the processor during a series of MAC operations, dominant in the GFDM kernel. The graph points out where the quantization effects act as a source of noise active upon the data in computation. The addition of quantization noise is a consequence of quantizing the "relative" data value[16,17] to a finite number of n + 1 (n fraction, 1 sign) bits. The reference as well as other parts of the test bed are coded in double-complex precision ($imag\,64-bit$, $real\,64-bit$); hence there is also type casting happening at kernel interfaces.

16 At each stage, quantization occurs anew, for example $0.2567 \xrightarrow[{[-1,1)}]{8\,bits} 0.2500 = 32_{Q0.7\,decimal}$.

17 Accumulator quantization: the result of multiplying Q0.n numbers is Q0.2n, which still represents a value in the range $[-1, 1)$, just with 2n bits, where $n = data_{bits}$. This value is then again requantized with $ACC_{bits} = m$ bits.

GFDM parameters are set to $K=512$ and $M=7$. These parameters can be extrapolated from section 1.2.2. K can be aligned with frequency (IDFT bin size for subcarriers) and M with time (symbols in a slot), for a representative 4G/5G compliant case. The *error vector magnitude* (EVM) budget is taken from standard specification (3GPP 2018b, 2019a), out of which 1% can be used for *digital baseband* (DBB) processing according to the engineering rule of thumb (Rohde & Schwarz *et al.* 2013). In these cases, the channel model is omitted. We measure the achieved EVM indirectly, by measuring signal and noise power and computing EVM, as shown in [1.2]

$$\text{EVM}_{\%} = \sqrt{\frac{P_{noise}}{P_{signal}}} \cdot 100\% \quad \text{EVM}_{dB} = 10log_{10}\frac{P_{noise}}{P_{signal}} \qquad [1.2]$$

where P_{noise} is the average error power of data carrying subcarriers and P_{signal} is the average power of data carrying subcarriers. The average power of data carrying subcarriers is invariant to QAM modulations; hence the EVM measured is the same for different QAM modulations, although the budget available is not the same. Measurement results are shown in Figure 1.12 in EVM_{dB} notation with the budget constraints overlapped over the results.

First, based on a theoretical bound for uniform quantization (Proakis and Manolakis 1996), as expected, we observe a decrease of EVM by 6 dB per bit of data bit-length in the constant region where the accumulator bit-length is sufficiently high. *Second*, when the accumulator bit-length is insufficient, we observe a signal quality degradation reaching the tangent (see equation [1.3])

$$\text{EVM}_{tangent}\,[dB] = -6.04 \cdot \text{ACC}_{bits} + 13.33 \qquad [1.3]$$

Third, the transition region is centered around the point $ACC_{bits}=data_{bits}$. The transition region extends $\lceil log_2 M \rceil$ (rounded towards infinity as in $\lceil 3.14 \rceil = 4$) bits away from the region center point. M corresponds to the number of MACs per accumulator of the algorithm shown in Figure 1.9, explaining the need for extra bits logarithmically proportional to the number of MACs. Hence, the following conclusions can be drawn:

– data type with a bit-length of 16 is sufficient for all standard required QAM modulations with a buffer budget ranging from $25.27\,dB$ for QPSK to $8.37\,dB$ for 1024-QAM. This leftover budget can be used in other DBB transmitter processing blocks;

– accumulator guard bits of at least $\lceil log_2 M \rceil$ are needed to avoid signal degradation due to the EVM transition region.

Since many SotA general-purpose DSPs today have an accumulator complex bit-length of 40 and data word length of 16, we conclude that there is no degradation of signal quality if we implement GFDM on a SotA DSP core. Other data types from the blue triangle (Figure 1.12) may be used as well to achieve further optimization of area and power, for example, the ASIP technology, while still meeting requirements for numerical precision. For generality, in the next section, we will keep ACC_{bits}, $data_{bits}$ as parameters.

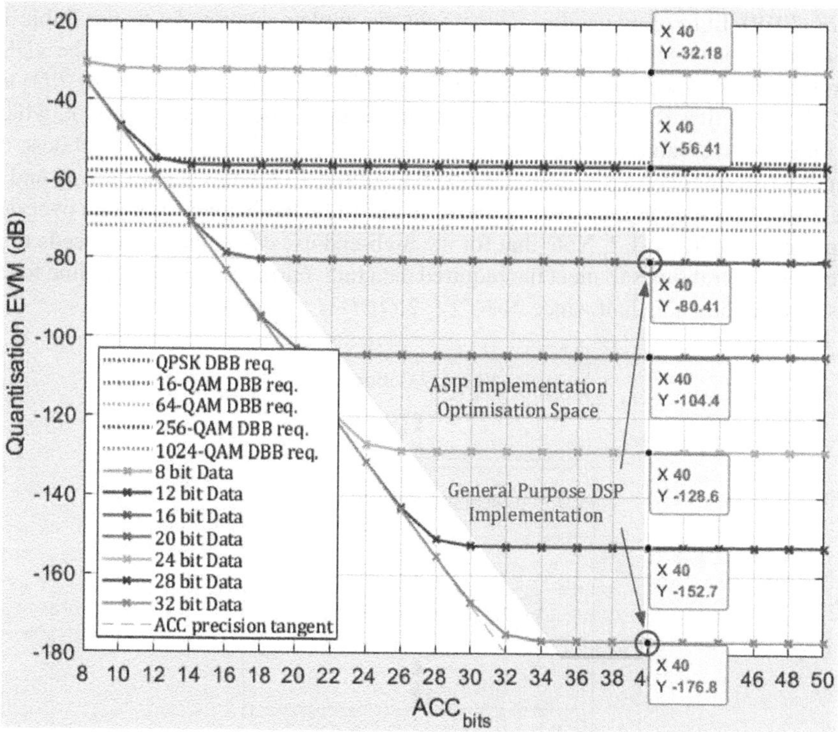

Figure 1.12. *GFDM EVM for varied data and ACC complex bit-lengths compared to adjusted 3GPP EVM DBB requirements (3GPP 2018b, 2019a)*

1.5. Implementation

Implementation of algorithms on wide vector processors, such as the vDSP that we used, introduces a series of considerations that need to be taken into account. Furthermore, the design implementation solution space increases even further by having algorithms with multiple loops. In the solution space, we need to make choices, for example, which loop should be vectorized or which loop order is notably impactful on the kernel's overall performance and requirements. These considerations add yet another layer of complexity and deserve a chapter of their own. Here, we will separate most important elements in an abridged manner to reach the next set of HW requirements: how does the 6G candidate waveform kernel under corner workloads map onto SotA vDSPs? How much of the vDSP core cycle budget does it require?

Is it practical to run the kernel on the vDSP, provided that the current vDSP load is sufficiently low?

A simplified block diagram of the vDSP is illustrated in Figure 1.13 (Damjancevic *et al.* 2019). But before we dive deeper into the implementation, let us use Table 1.1 to identify for which set of GFDM kernel parameters we need to profile the vDSP. As mentioned in section 1.4, the subcarriers $(RE/(Symbol \cdot BW \cdot layer \cdot TTI))$ use an IDFT bin that is slightly larger, i.e. rounded up to the nearest step of 2, which, in turn, sets the value for K, as in $K = bin\, size$. On the other hand, we choose the number of symbols per slot to be M, i.e. half of the TTI. Other values for M and K are possible as long as they overlap and fit in the TTI in both dimensions. An overview is presented in Table 1.2. Note that for the high-end use cases, the kernel needs to be executed several times to meet the required data rate. Furthermore, the deadline for all cases scales down by half, since $M \hat{=} TTI/2\,[OFDM\ Symbols]$.

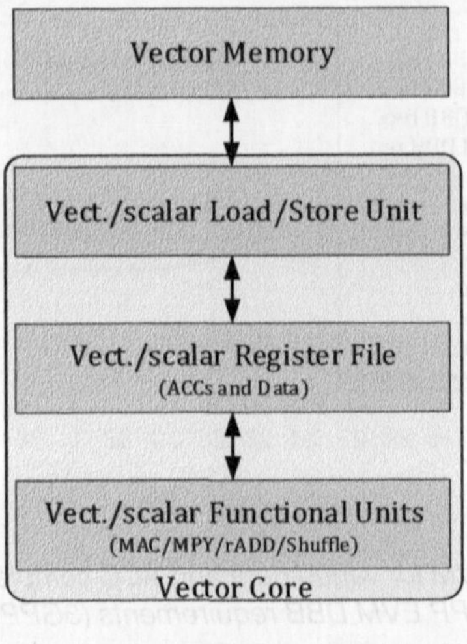

Figure 1.13. *vDSP simplified HW block diagram*

Use Case	Throughput $\left[\frac{RE}{symbol \cdot BW \cdot layer \cdot TTI}\right]$	TTI $[\mu s]$	K [#]	M [#]	Kernels req. [#]	Deadline $[\mu s]$
low-end LTE legacy	72	1000	128	7	1	500
high-end FR2 $4 \times CA, \mu = 3, 400MHz$	3,168	125	4,096	7	4	62.5
MIMO high-end FR2 $8 \times 8, 4 \times CA, \mu = 3,400MHz$	3,168	125	4,096	7	32	62.5

Table 1.2. *Kernel parameters for corner use cases*

1.5.1. *Implementation considerations*

For the kernel implementation, we need to factor in the following shortened list of considerations (Damjancevic *et al.* 2019):

1) *Vectorization* – In an ideal case, single-instruction, multiple-data (SIMD) processors of vector length vec_{len}, i.e. vec_{len} SIMD-lanes, will reduce the number of operations required by a factor of vec_{len}, but, in practice, this number is smaller, due to Amdahl's law, boundary conditions and other potential bottlenecks, such as load/store bandwidth. Similarly, the number of memory accesses (loads/stores) in the ideal case is reduced vec_{len} times.

2) *Boundary conditions* – Based on the algorithm implemented, the mismatch and misalignment of data in memory and data accessed by the vector load/store unit (Figure 1.13) lead to a need for extra processing cycles.

3) *Bandwidth bottleneck* – When the functional units and register file blocks of Figure 1.13 are consuming or producing more data than the load/store block can handle, the core will run into a BW bottleneck, lowering the core utilization. We could use multiple load/store units and/or increased load/store vector length to mitigate this problem, although such architectures come with higher energy consumption per load/store and a bigger area on the chip needed for the load/store unit.

4) *Memory bandwidth* – Amount of data transferred per unit of time.

5) *Trade-off cycle count and memory bandwidth* – Sometimes, when implementing an algorithm, it would be beneficial to minimize cycle counts, which is usually paid for in additional memory accesses and vice versa, optimizing for a low load/store count would increase the overall cycle count. In GFDM, we encounter such a trade-off. Low cycle count optimization reduces execution time, and a low memory bandwidth optimization saves energy (and area, if using an architecture with a smaller load/store unit).

6) *Loop order and vectorization* – Some algorithms, such as GFDM, can be described with mutually independent nested loops. We identified three such loops in GFDM (see "for" loops in Figure 1.9). Loop order and choosing which loop to vectorize can significantly alter the boundary conditions, bandwidth, memory and register space used and even type (*instruction set architecture* (ISA)) and count of operations needed.

1.5.2. *Design space exploration*

Let us refer to Figure 1.8 again. Note that the matrix elements are color-coded to match outputs with inputs.

Three loops can be identified: loop i along K, loop m along M of the dot product and loop l along M of the cyclic shift. We will use this notation for the rest of this chapter. $\boldsymbol{D}$ matrix is m and i dependent; $\boldsymbol{g}$ matrix is l, m and i dependent and output $\boldsymbol{x}$ is i and l dependent. There are 6 (3!) loop order combinations to arrange data reading, computation and data storing in this algorithm. In addition, we can vectorize loops i or l at the output and any of the loops at the input. In our notation, a vectorized input loop is emboldened, for example, $i \rightarrow \boldsymbol{i}$, and a vectorized output loop is underlined, for example, $i \rightarrow \underline{i}$.

In the design space, we can search for optimizations in regard to section 1.5.1. We decided to search for two optimizations: a) ideal vectorization and b) low memory bandwidth. In a), we analyze two variants (black and red) and in b) two other variants (blue and green) of GFDM. Due to some features overlapping between optimization variants, in the following text we will first present features and then assign them accordingly to black, red, blue and green variants. A summary is provided in Table 1.3 (Damjancevic *et al.* 2019). The features are:

1) *Modifying the very long instruction word (VLIW) configuration* – The vDSP has the capability to perform either a wide $2vec_{len}$ vector load or a vec_{len} vector load + vec_{len} vector store. This removes the *bandwidth bottleneck* for operations that require two operands, which, in turn, removes stalls when executing consecutive vector MAC operations. Basic configuration supports a vec_{len} vector load + vec_{len} vector store.

2) *Vectorizing loop i maximizes efficiency (no boundary effects)* – The IDFT output is of size $K = 2^i$, $i \in \mathbb{N}$ due to the assumed algorithm implementation of IDFT as 2^i point iFFT ($K \in [128, 4096]$ *i.e.[low end, high end]*), and vec_{len}=16. Moreover, this makes the code scale with any M and K value.

3) *Vectorizing loop m* allows the use of a) a *shuffle* operation on buffered coefficients $\boldsymbol{g}$, reducing the required number of $\boldsymbol{g}$ loads from M^2K to MK; b) a combination of element-wise *vector multiplication* (MPY) and *reduction add* (rADD) that avoids MACs, and if $M \leq vec_{len}$, then the m loop is consumed altogether. This greatly reduces the number of memory accesses. Adversely, for the cases when M is not a composite of vec_{len}, the boundary conditions deteriorate and the efficiency drops. We could handcraft the code for every M to optimize boundary effects, but this is a limitation to code scalability. In Table 1.3, we present the upper bound of efficiency for arbitrary M.

4) *Vectorizing loop l at the output* reduces the operation count, but is subject to *boundary effects* that reduce core efficiency.

5) *m nested in i* minimizes the need for ACCs in the register file (see Figure 1.13). Variants with this loop order use 1 vec_{len} ACC, by computing the partial sums of the dot product first.

Algorithm	Black	Red	Blue	Green
Loop order, vectorisation	$l\{\underline{i}\{m\{\ldots\}\}\}$	$m\{\underline{\mathbf{i}}\{l\{\ldots\}\}\}$	$i\{\underline{l}\{\mathbf{m}\{\ldots\}\}\}$	$l\{\underline{i}\{\mathbf{m}\{\ldots\}\}\}$
Vectorisation efficiency η (upper bound)	1		$1-\frac{vec_{len} modM}{vec_{len}}$ if $vec_{len} \geq M$; $1-\frac{(nvec_{len}) modM}{nvec_{len}} \mid n \in \mathbb{N}, nvec_{len} \in [M, 2M)$ if $vec_{len} < M$	
Min. scalar operations	$M^2K + \left(2M^2K + MK\right)$ [arithm./logical + (loads + stores ≡ memory accesses)]			
Min. vector ops. (arith./log. + mem.accesses)	$\frac{M^2K}{vec_{len}}+$ mem.acc.		$\frac{M^2K + floor\left(\frac{vec_{len}}{M}\right)MK}{vec_{len}\eta} + \frac{MK}{vec_{len}} + \frac{MK}{vec_{len}\eta}$ +mem.acc.	$\frac{M^2K + floor\left(\frac{vec_{len}}{M}\right)MK}{vec_{len}\eta} + \frac{2MK}{vec_{len}}$ +mem.acc.
Required ISA	MAC, cyclic load		MPY, rADD, Shuffle	
Used VLIW config.	$\frac{2\,loads}{cycle}$ and $\frac{load + store}{cycle}$		$\frac{load + store}{cycle}$	
Memory accesses	$\frac{2M^2K}{vec_{len}}$ loads $\frac{MK}{vec_{len}}$ stores	$\frac{MK+M^2K}{vec_{len}}$ loads $\frac{MK}{vec_{len}}$ stores	$\frac{2MK}{vec_{len}\eta}$ loads $\frac{MK}{vec_{len}\eta}$ stores	$\frac{2MK}{vec_{len}\eta}$ loads $\frac{MK}{vec_{len}}$ stores
Memory (in, out) [complex bits]	$data_{bits} \times (2MK, K)$	$data_{bits} \times (K + MK, vec_{len}K)$	$data_{bits} \times (2MK, 0)$	$data_{bits} \times (2MK, K)$
Register file [complex bits]	$data_{bits} \times 3vec_{len} + ACC_{bits} \times vec_{len}$	$2data_{bits} \times 3vec_{len} + ACC_{bits} \times MK$	$data_{bits} \times (4vec_{len} + 1) + ACC_{bits} \times (vec_{len} + 1)$	$data_{bits} \times \left(4vec_{len} + \frac{2MK}{vec_{len}\eta} + 1\right) + ACC_{bits} \times (vec_{len} + 1)$
Code scales	no need to reoptimise for any M or K		K scales, but for every M the code needs to be reoptimised	

Table 1.3. *Selected GFDM implementation variants*

6) *i nested in l* produces the output sequentially on a symbol-by-symbol basis of size K.

7) *i nested in m* allows ***D*** to be read sequentially symbol-by-symbol, preventing stalls when the IFFT result is incoming sequentially (blocks of size K in contrast to the otherwise parallel $M \times K$ blocks requirement).

8) *l nested in i* removes the need to either reload or buffer ***D*** in register file.

9) *m and l nested in i* allow the use of the same memory locations for both ***D*** and ***x***. An index location of ***D*** becomes obsolete when ***x*** of the same index location is ready, for example, when $x_{0,0}$ is ready, $D_{0,0}$ is no longer used. The relative order of m and l is irrelevant. The same is true for i nested in m and l.

Black uses $l\,\{\underline{i}\,\{m\,\{\ldots\}\}\}$, hence 1), 2), 5), 6) are valid.

Red uses $m\,\{\underline{i}\,\{l\,\{\ldots\}\}\}$, hence 1), 2), 7), 8) are valid.

Blue uses $i\,\{\underline{l}\,\{\boldsymbol{m}\,\{\ldots\}\}\}$, hence 3), 5), 8), 9) are valid.

Green uses $l\,\{\underline{i}\,\{\boldsymbol{m}\,\{\ldots\}\}\}$, hence 3), 5), 6) are valid. In green, we have to either reload or buffer ***D***; we chose buffer ***D*** in register file, reducing memory bandwidth.

Now that we have identified GFDM variants to minimize cycle count (for maximum throughput) or required memory bandwidth (for low energy consumption), we can investigate how the variable throughput requirement from the standard analysis in section 1.1 maps to GFDM kernels, in particular, to black – high-throughput variant, and blue – low-energy-consumption variant. We drop red and green from further consideration since they do not achieve the lowest cycle counts and have high-memory register file usage. However, we should keep in mind the lesson learned: that our code could create a kernel that is in a very unfavorable spot on the design space, if we were to underestimate all the complexity layers and their implications.

1.5.3. *Measurements for low-end and high-end use cases*

For the analysis of GFDM kernel performance for the high-end and low-end workloads posed by the standard analysis in section 1.2 and Table 1.1, we have to provide a vDSP architecture on which to execute the GFDM kernel.

Referring to Figure 1.13, we designed a cycle-accurate simulation model of the vDSP in the ASIP Designer tool suite (Synopsys Inc 2019). The tool comes with a C-compiler and a debugging and profiling environment, which allowed for practical measurements of realistic cycle counts for the GFDM variants implementation. In a reiterative process, we have recursively tuned the prototype, refining the HW architecture instruction set and VLIW configuration, editing functional units of the vDSP, parallel to the development of the black and blue vector codes. We have arrived at an adequate processor and corresponding black and blue optimized code variants. Based on considerations thus far, we define 512 bit SIMD (vector) processing, 16 bit complex data types, 40 bit complex accumulators and 1 GHz clock.

We use theoretical minimum operations as the baseline for a fair, objective and unbiased comparison between the code variants and their utilization of processor resources. We call this metric *implementation execution density* ρ and define it as a ratio of minimum theoretical operations and measured cycles. The theoretical minimum includes a) only general/standard arithmetic or logical operations (not fused/composite operations that combine several into one, with MAC as the only exception), and b) memory accesses. The theoretical vector operations minimum depends on the implementation variant, i.e. the loop order combination and vectorization.

Table 1.4 gives us the GFDM kernel profile. The variant-specific minimum operations are taken from Table 1.3 for the ρ calculation. From Table 1.4, we can note a) the difference in total memory accesses and cycle counts between the two code variants as a consequence of different loop ordering and loop vectorization; b) even $\rho_v > 1$ values are possible when optimizing the low cycle count (black) variant through thoughtful code scheduling on the processor VLIW architecture with multiple instruction issue slots, with 512 bit SIMD capability per slot, and efficient pipelining for the selected loop order and vectorization; and c) ρ_s shows the ratio between the theoretical best case scalar execution and our kernel and can be interpreted as a *speed up* against the optimal theoretical scalar processor implementation.

Kernel Parameters		Metric	Black	Blue
$M = 7$	$K = 128$	$cycles$	504	2,695
		$loads, stores$	$784, 56$	$128, 64$
		$\rho_v \left(\frac{min.\ vect.\ ops}{cycles}\right)$	$2.44\ (\frac{1232}{504})$	$0.33\ \left(\frac{888}{2{,}695}\right)$
		$\rho_s \left(\frac{min.\ scalar\ ops}{cycles}\right)$	$39.11\ (\frac{19{,}712}{504})$	$7.31\ \left(\frac{19{,}712}{2{,}695}\right)$
	$K = 4{,}096$	$cycles$	14394	86,023
		$loads, stores$	$25088, 1792$	$4096, 2048$
		$\rho_v \left(\frac{min.\ vect.\ ops}{cycles}\right)$	$2.73\ \left(\frac{39{,}424}{14{,}394}\right)$	$0.33\ \left(\frac{28{,}416}{86{,}023}\right)$
		$\rho_s \left(\frac{min.\ scalar\ ops}{cycles}\right)$	$43.82\ \left(\frac{630{,}784}{14{,}394}\right)$	$7.33\ \left(\frac{630{,}784}{86{,}023}\right)$

Table 1.4. *Kernel profile: cycles, memory accesses, and density*

As the last step in our investigation, we show the required frequency budget needed for the corner cases, in Table 1.5. To form the budget requirements, we need to take into account the cycle counts from Table 1.4, as well as number of kernel calls, kernel parameters and deadlines from Table 1.2. Finally, we show the mentioned budget requirement in the context of our vDSP. In particular, how large a portion of the total vDSP frequency budget is needed to meet the specific use case requirements. Higher than 100% usage means that more than 1 vDSP core is needed.

Use Case	Metric		Black	Blue
low-end LTE legacy $200\,kb/s \equiv \frac{1\,call}{500\,\mu s} \cdot kernel_{128}$	$required\,budget\ [MHz]$		1.01	5.39
	$our\ vDSP$	$processing\,time\ [\mu s]$	0.504	2.695
		$vDSP\,utilization\ [\%]$	0.10	0.54
		$min.\,vDSPs\,to\,run\ [\#]$	1	1
CA high-end FR2 $10.52\,Gb/s \equiv \frac{4\,calls}{62.5\,\mu s} \cdot kernel_{4096}$ $4{\times}CA, \mu = 3{,}400MHz$	$required\,budget\ [MHz]$		921.2	5,505.5
	$our\ vDSP$	$processing\,time\ [\mu s]$	57.58	344.09
		$vDSP\,utilization\ [\%]$	92.12	550.55
		$min.\,vDSPs\,to\,run\ [\#]$	1	6
MIMO CA high-end FR2 $84.16\,Gb/s \equiv \frac{32\,calls}{62.5\,\mu s} \cdot kernel_{4096}$ $8{\times}8, 4{\times}CA, \mu = 3{,}400MHz$	$required\,budget\ [MHz]$		7.37	44.04
	$our\ vDSP$	$processing\,time\ [\mu s]$	460.6	2,752.7
		$vDSP\,utilization\ [\%]$	736.95	4,404.38
		$min.\,vDSPs\,to\,run\ [\#]$	8	45

Table 1.5. *GFDM: required frequency budget and performance on our vDSP*

The results argue in favor of the discussion from section 1.2: it is practical to run the low-end use cases quite effortlessly in parallel with other kernels and tasks on a vDSP. Surprisingly, even the CA high-end black GFDM can fit on a single vDSP core and make the deadline, albeit at a heavy load. Since there are several vDSPs on the MPSoC, running this modulation flavor is an option to consider, provided, of course, that the memory bandwidth allows using the black flavor. Finally, as expected, it is practical to use HW accelerator engines for the MIMO CA high-end use case instead of many fully loaded vDSP cores.

1.6. Conclusion

This chapter closely followed an SW implementation of the GFDM algorithm on the SotA vDSP and noted considerations taken into account with regard to handset workloads expected in modern and future mobile communications. We give analyses and conclusions on four layers: specification requirements, translating theory to pseudo-code, precision analysis and requirements and implementation space exploration.

First, we indicated trends and analyzed specifications for the most advanced 5G communication standard at the time of writing this chapter and defined workloads based on these specifications. To get closer to HW requirements, we selected an example beyond 5G algorithm GFDM to implement and investigate its HW requirements for those workloads. Second, from the theory behind GFDM, we developed the processing graph and pseudo-code that does not use modulo operations. Third, the precision analysis indicated that HW should be able to handle data types of at least 32 bits (16 bit – real, 16 bit – imag) per data point for operation and that the processor accumulators should have at least $\lceil log_2 M \rceil$ guard bits to facilitate computations without degradation. Fourth, we identified the range of kernel parameters as well as the number of calls and deadlines for the low- and high-end 5G workloads, followed by the implementation space exploration, resulting in high-throughput and low-memory access code variants of the same kernel. Finally, the kernel variants and workloads are combined to indicate required HW for corner cases. For the low-end and CA high-end vDSPs are recommended, while for higher than those and towards MIMO CA high-end workloads, more specialized (application-specific) HW accelerators or ASIPs are recommended.

We can see that giving a recommendation or direct HW requirements for SotA specifications is a multi-layered challenge that is difficult to overcome. For the overall system, other kernels can be analyzed in the same manner; however, the overarching theme is the need for variability and flexibility at each step of the analysis. Heterogeneous HW MPSoCs consisting of RISCs, vDSPs, ASIPs and dedicated (application-specific) HW accelerators can provide both flexibility and highly optimized solutions in the power consumption/chip area/chip cost sense, provided that enough care and effort has been invested into the investigation of the above-mentioned layers.

1.7. Acknowledgments

The authors would like to thank Dr. Yankin Tanurhan and the Synopsys team for their guidance, support, sponsorship and initiation of the Industry–University cooperation within the “Efficient Implementation of 5G Baseband Kernels on a Vector Processor” project.

1.8. References

3GPP (2017). Study on new radio access technology Physical layer aspects v14.2.0, TR 38.802 [Online]. Available at: https://bit.ly/2QzxmGc.

3GPP (2018a). Evolved universal terrestrial radio access (E-UTRA); Physical channels and modulation v15.3.0, TS 36.211, September 2018 [Online]. Available at: https://bit.ly/354hKjf.

3GPP (2018b). NR; Base Station (BS) radio transmission and reception v15.3.0, TS 38.104, September 2018 [Online]. Available at: https://bit.ly/30xf5wE.

3GPP (2018c). NR; Physical channels and modulation v15.0.0, TS 38.211, January 2018 [Online]. Available at: https://bit.ly/2xO4Qrt.

3GPP (2018d). Study on scenarios and requirements for next generation access technologies v15.0.0, TR 38.913 [Online]. Available at: https://bit.ly/2MCblVZ.

3GPP (2019a). Evolved universal terrestrial radio access (E-UTRA); base station (BS) radio transmission and reception v16.3.0, TS 36.104 [Online]. Available at: https://bit.ly/2xLJSto.

3GPP (2019b). Evolved universal terrestrial radio access (E-UTRA); Physical layer procedures v15.6.0, TS 36.213 [Online]. Available at: https://bit.ly/2SgRopA.

3GPP (2019c). NR; User equipment (UE) radio transmission and reception; Part 1: Range 1 standalone v15.7.0, TS 38.101-1 [Online]. Available at: https://bit.ly/2LUtz5K.

3GPP (2019d). NR; User equipment (UE) radio transmission and reception; Part 2: Range 2 standalone v15.7.0, TS 38.101-2 [Online]. Available at: https://bit.ly/2SfjQIc.

3GPP (2019e). NR; Physical channels and modulation v15.7.0, TS 38.211, September 2019 [Online]. Available at: https://bit.ly/2xO4Qrt.

3GPP (2019f). NR; Physical layer procedures for data v15.4.0, TS 38.214 [Online]. Available at: https://bit.ly/2Gg9aVb.

ARIB, ATIS, CCSA, ETSI, TSDSI, TTA, and TTC (2016). Study on new services and markets technology enablers; Stage 1 v14.2.0, TR 22.891 [Online]. Available at: https://bit.ly/39meeUM.

Damjancevic, S.A., Matus, E., Utyansky, D., van der Wolf, P., and Fettweis, G.P. (2019). Towards GFDM for handsets - Efficient and scalable implementation on a vector DSP. *IEEE Vehicular Technology Conference (VTC Fall 2019)*, 1–7.

Farhang, A., Marchetti, N., and Doyle, L.E. (2016). Low-complexity modem design for GFDM. *IEEE Transactions on Signal Processing*, 64(6), 1507–1518.

Fettweis, G.P. (2012). A 5G wireless communications vision. *Microwave Journal*, 55, 24–36 [Online]. Available at: https://bit.ly/368uSW7.

Fettweis, G.P. (2014). The tactile Internet. *IEEE Vehicular Technology Magazine*, 9(1), 64–70.

Fettweis, G.P., Hassler, M., Wittig, R., Matus, E., Damjancevic, S., Haas, S., Pauls, F., Nam, S., and Grigoryan, N. (2019). A low-power scalable signal processing chip platform for 5G and beyond – Kachel. *2019 53rd Asilomar Conference on Signals, Systems, and Computers*.

GFK (2019). Global smartphone sales reached $522 billion in 2018 [Online]. Available at: https://www.gfk.com/insights/press-release/global-smartphone-sales-reached-522-billion-in-2018/ [Accessed 26 December 2019].

ITU (2019a). G.722 : 7 kHz audio-coding within 64 kbit/s [Online]. Available at: https://www.itu.int/rec/T-REC-G.722/e [Accessed 4 August 2020].

ITU (2019b). G.729 : Coding of speech at 8 kbit/s using conjugate-structure algebraic-code-excited linear prediction (CS-ACELP) [Online]. Available at: https://www.itu.int/rec/T-REC-G.729 [Accessed 4 August 2020].

Matthé, M., Mendes, L., Gaspar, I., Michailow, N., Zhang, D., and Fettweis, G.P. (2016). Precoded GFDM transceiver with low complexity time domain processing. *EURASIP Journal on Wireless Communications and Networking*, 2016(138), 1–9. [Online]. Available at: http://dx.doi.org/10.1186/s13638-016-0633-1.

Michailow, N., Matthé, M., Caldevilla, A.N., Mendes, L.L., Festag, A., and Fettweis, G.P. (2014). Generalized frequency division multiplexing for 5th generation cellular networks. *IEEE Transactions on Communications*, 62(9), 3045–3060.

NGMN Alliance (2015). NGMN 5G white paper [Online]. Available at: https://www.ngmn.org/wp-content/uploads/NGMN_5G_White_Paper_V1_0.pdf.

Nimr, A., Chafii, M., and Fettweis, G.P. (2018). Unified low complexity Radix-2 architectures for time and frequency-domain GFDM modem. *IEEE Circuits and Systems Magazine*, 18(4), 18–31.

Proakis, J.G. and Manolakis, D.G. (1996). *Digital Signal Processing: Principles, Algorithms, and Applications*, 3rd edition. Prentice-Hall, Upper Saddle River.

Qualcomm (2016). Making 5G NR a reality. Leading the technology innovations for a unified, more capable 5G air interface. *CTIA Super Mobility 2016 – 5G Technical Workshop*.

Synopsys Inc (2019). ASIP designer [Online]. Available at: https://www.synopsys.com/dw/ipdir.php?ds=asip-designer [Accessed 4 August 2020].

Werther, O., and Minihold, R. (2013). LTE: System specifications and their impact on RF & base band circuits. Application note, Rohde & Schwarz.

2

Towards Tbit/s Wireless Communication Baseband Processing: When Shannon meets Moore

Matthias HERRMANN and Norbert WEHN

Microelectronic Systems Design Research Group, TU Kaiserslautern, Germany

Mobile communication plays a central role in our information society and is a key enabler in our connected world. The newest standard, 5G, features data rates >10 Gbit/s. Beyond 5G, data rates towards 1 Tbit/s are expected. The tremendous improvement in mobile communication has to be considered alongside the progress in the microelectronic industry. For many decades, improvement in silicon process technology provided better performance, lower cost per gate, higher integration density and lower power consumption. However, we have reached a point where Moore's law is slowing down and microelectronics cannot keep up with the increased requirements coming from communication systems. Thus, the design of communication systems is no longer just a matter of spectral efficiency or bit/frame error rate. When it comes to implementation, design of communication systems requires a cross-layer approach covering information theory, algorithms, parallel hardware architectures and semiconductor technology to achieve excellent communications performance and high implementation efficiency. In this chapter, we focus on channel coding, which is a major source of complexity in digital baseband processing, and will highlight implementation challenges for the most advanced

For a color version of all figures in this book, see www.iste.co.uk/andrade/multi2.zip.

Multi-Processor System-on-Chip 2 – Applications,
coordinated by Liliana ANDRADE and Frédéric ROUSSEAU.

channel coding techniques, i.e. Turbo codes, Low Density Parity Check (LDPC) codes and Polar codes, for throughput requirements towards 1 Tbit/s.

2.1. Introduction

Wireless communication plays a central role in our information society and is a key enabler for our connected world. Today we already have more than 20 billion connected devices, and the majority of them are wirelessly connected. The first generation of mobile communication systems has shifted the communication from landline to handheld devices, followed by the third (3G) and fourth generations (4G) that marked the advent of the mobile internet. We have seen a tremendous increase in data rates over the different generations, for example, the Global System for Mobile Communications (GSM) featured about 10 kbps, the Universal Mobile Telecommunications System (UMTS) about 2 Mbit/s and Long Term Evolution Advanced (LTE-A) about 1 Gbit/s. The newest standard, 5G, features data rates >10 Gbit/s. Beyond 5G, data rates toward 1 Tbit/s are expected (EPIC 2020). However, increasing data rates are not the only driver. With the advent of 5G, wireless communication is finding its way into a variety of applications with largely diverse requirements. Massive Internet of Things (IoT) demands for extremely low power, massive content applications (e.g. video streaming, cloud office) call for very high throughput and the use of wireless communication in control applications (e.g. autonomous driving, remote control, tactile internet) requires very fast response times in the order of 1 ms. Thus, communication has to support a huge throughput–latency range (5 orders of magnitude in throughput, 3 orders of magnitude in latency) and a large diversity of applications with different Bit Error Rate (BER)/Frame Error Rate (FER) requirements (Fettweis and Matus 2017). In this chapter, we focus on the digital baseband processing part of wireless communication. The main tasks of baseband processing are signal detection, parameter estimation, demodulation and channel decoding. Figure 2.1 shows a state-of-the-art commercial advanced baseband System-on-Chip (SoC) from the company Octasic (*http://www.octasic.com*). This SoC supports a wide range of communication standards from 2G to 5G. Assisting different standards demands for flexibility; on the other hand, power and energy efficiency requires dedicated optimized accelerators that demand a careful flexibility/implementation efficiency trade-off. A specialized hardware is 2 to 3 orders of magnitude more efficient than processor-based solutions that offer largest flexibility (Horowitz 2014). Hence, the SoC features a heterogeneous architecture composed of standard CPUs such as ARM cores, highly optimized low-power Digital Signal Processor (DSP) cores and dedicated accelerators.

The most computational-intensive part in baseband processing is channel decoding. Channel coding is an essential part in any communication system to guarantee reliable transmission, and enables the correction of errors that are induced during data transmission over noisy channels. It has a long history, going back to Shannon's famous channel coding theorem in 1948 (Shannon 1948), and is based on the solid theoretical mathematical framework of information theory. Coding evolved as a very active research field and achieved a major breakthrough in 1993, when Claude Berrou wondered whether two small convolutional codes arranged in an original way, and iteratively decoded one after another, could do better than convolutional codes, which are based on very large states. This new coding scheme was named Turbo codes (Berrou *et al.* 1993). The era of iterative coding schemes was born, which resulted in a re-discovery of LDPC codes by MacKay and Neal (1997) that originally had already been published by Gallager in 1962 (Gallager 1962). Due to their excellent communications performance, consumer-driven cellular and broadcast communication standards such as UMTS, LTE, DVB and WiMAX, to name a few, adopted these new codes. A further breakthrough happened in 2009 when Erdal Arikan published Polar codes and proved that these codes could achieve channel capacity for a Binary Symmetric Memoryless Channel Binary Symmetric Memoryless Channel (BSMC) (Arikan 2009). Due to their excellent communications performance, Polar codes have attracted significant attention and became, for the first time, part of the 5G standard. For high throughput and low latency, these advanced decoding schemes, however, imply large challenges for efficient hardware implementation in consumer devices, where silicon area and power are the most critical cost factors. The computational workload of channel decoding for these advanced coding schemes with throughput in the Gbit/s range is more than 1,000 GOP/s (equivalent DSP operations). It is even more challenging with regard to the available power budget that is limited to some Watts for baseband processing due to thermal power density constraints. The high design costs of a baseband chip in advanced semiconductor technology demands for a large volume. This means that the corresponding SoC has to be used in various applications and the Forward Error Correction (FEC) Intellectual Property (IP) must be flexible to support various coding schemes, code rates, data rates, latency and BER/FER requirements. As mentioned earlier, adding flexibility to any architecture has a negative impact on throughput, energy efficiency, and area. Researchers are continuously looking for techniques to reduce the hardware implementation complexity and optimize energy efficiency for these advanced coding schemes while minimizing the degradation in communications performance (Kienle *et al.* 2011). Because of the aforementioned reasons, state-of-the-art channel decoders are typically implemented as dedicated accelerator cores. This can be seen in the wireless accelerator group depicted in Figure 2.1, that contains different types of channel en-/decoding accelerators to support various standards: an Application-Specific Instruction Set Processor (ASIP) that offers large flexibility for 2G/3G convolutional and Turbo code support (Vogt and Wehn 2008) and dedicated highly optimized LTE Turbo and DVB-S2 LDPC en-/decoders (Muller *et al.* 2009; Weithoffer *et al.* 2018).

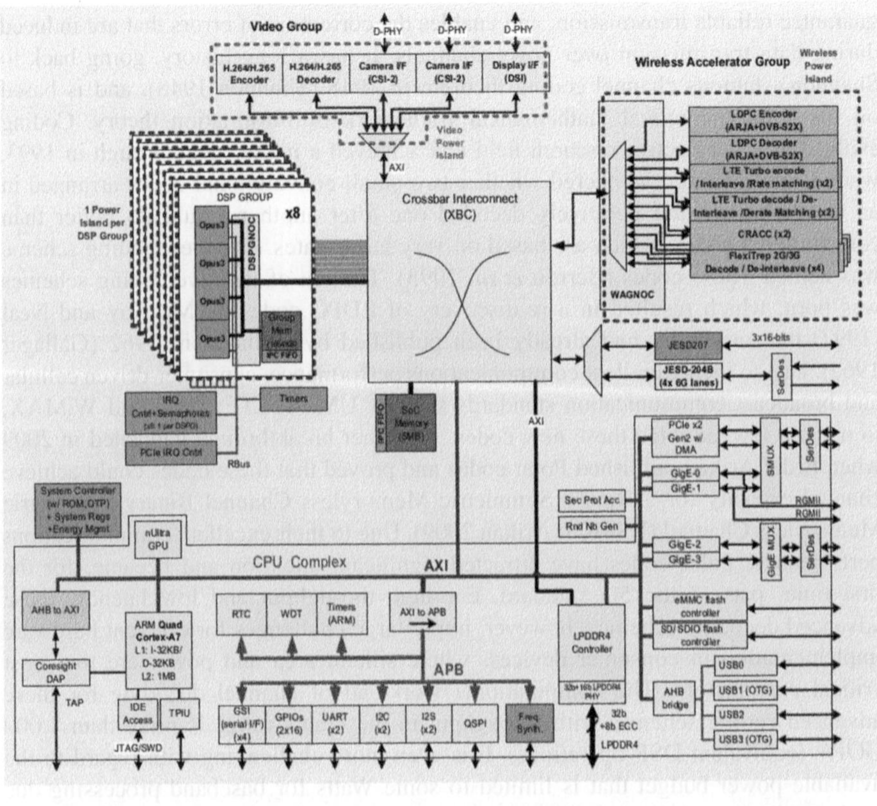

Figure 2.1. *State-of-the-art commercial system-on-chip baseband architecture*

2.2. Role of microelectronics

The tremendous improvement in mobile communication has to be considered alongside the progress in the microelectronic industry, which started with the invention of the transistor in the late 1940s (Shockley 1949), coincidentally at the same time as, when Shannon published his famous article (Shannon 1948). In the following decades, the semiconductor industry achieved an exponential increase in the number of transistors on a single chip, known as Moore's law (Moore 1965), which is a further key driver of our information society. In today's semiconductor technologies, two-digit million transistors can be integrated on 1 mm^2 of silicon. For many decades, improvement in the silicon process technology provided better performance, lower cost per gate, higher integration density and lower power consumption. However, we have reached a point where Moore's law is slowing down. The reasons for this slowdown are, in particular, the immense cost of new technologies and the design cost in these technologies, decreasing performance gain and increasing delay in interconnect and power/power density challenges, to name just a few.

The question is, what contribution have microelectronics made to improve throughput and implementation efficiency in channel decoding in the past. As a case study, we consider two Turbo code decoders. Both decoders were designed with the same design methodology and have a very similar state-of-the-art architecture that exploits spatial parallelism and processes several sub-blocks on corresponding Maximum a Posteriori (MAP) decoders in parallel:

– the first decoder is a fully UMTS-compliant Turbo decoder implemented in a $180\,\mathrm{nm}$ technology. Under worst-case Process, Voltage and Temperature (PVT) conditions, a maximum frequency of $166\,\mathrm{MHz}$ is achieved, which results in a throughput of $71\,\mathrm{Mbit/s}$ at 6 decoding iterations. The total area is $30\,\mathrm{mm}^2$ (Thul *et al.* 2005);

– the second decoder is a fully LTE-compliant Turbo decoder implemented in a $65\,\mathrm{nm}$ technology, achieving a maximum frequency of $450\,\mathrm{MHz}$ under worst-case PVT conditions. It yields a throughput of $2.15\,\mathrm{Gbit/s}$ at 6 decoding iterations and consumes $7.7\,\mathrm{mm}^2$ area (Ilnseher *et al.* 2012).

Three semiconductor technology nodes are between $180\,\mathrm{nm}$ and $65\,\mathrm{nm}$ technology. We observe a throughput increase by $30\times$ although the improvement of frequency, which is limited by the critical path inside the MAP decoder, is only $3\times$. The improvement in area efficiency (throughput/area) is $118\times$. Hence, progress in microelectronics contributed to a huge improvement in area efficiency but much less to a frequency increase, and, thus, throughput increase. The throughput increase mainly originates from code design, i.e. conflict-free Turbo code interleavers that enable efficient implementation with a high degree of parallelism, advanced algorithmic and architectural features, such as next-iteration initialization, optimized radix-4 kernel, re-computation and advanced normalization to reduce internal bit widths. We see that microelectronics could not keep up with the increased requirements coming from communication systems. Thus, the design of communication systems is no longer just a matter of spectral efficiency or BER/FER. When it comes to implementation, channel coding requires a cross-layer approach covering information theory, algorithms, parallel hardware architectures and semiconductor technology to achieve excellent communications performance, high throughput, low latency, low power consumption and high energy and area efficiency (Scholl *et al.* 2016; Kestel *et al.* 2018a).

2.3. Towards 1 Tbit/s throughput decoders

A large parallelism is a must for high-throughput decoders towards $1\,\mathrm{Tbit/s}$. The achievable parallelism strongly depends on the properties of the decoding algorithms, for example, sub-functions of a decoding algorithm that have no mutual data dependencies can easily be parallelized by spatial parallelism. This is the case for the Belief Propagation (BP) algorithm to decode LDPC codes. In the BP algorithm, all check nodes can be processed independently from each other. The same applies for the variable nodes. The situation is different for the MAP algorithm used in Turbo code decoding, where the calculation of a specific trellis step depends on the result of the previous trellis step. This results in a sequential behavior, and the different trellis steps cannot be calculated in parallel. Such data dependencies exist also in iterative decoding algorithms between the various iterations. Hence, the bottleneck for high throughput is the part of an algorithm that cannot be parallelized. This part strongly limits the overall throughput that is also known as Amdahl's law (Amdahl 2013). Other important features for efficient high-throughput implementations are locality and regularity to minimize interconnect. Interconnect can largely contribute to area, delay and energy consumption. Let us assume an FEC IP block of size $10\,mm^2$ is a feasible size. A signal has to travel at least 7 mm if it has to be transmitted from one corner to the other in this IP block. It will take three clock cycles in a 14 nm technology, assuming a frequency of 1 GHz. This interconnect delay largely decreases the throughput and increases power. Thus, data transfers can be as important as calculations. Although channel decoding algorithms for advanced codes such as Turbo- and LDPC codes are mainly data-flow dominated, they imply irregularity and restricted locality, since efficient channel coding for these coding schemes is grounded on some randomness: that is the interleaver for Turbo codes and the Tanner graph for LDPC codes, respectively. Any regularity and locality in these structures improves the implementation efficiency but has negative impact on the communications performance. Hence, there is a fundamental discrepancy between information theory and efficient implementations for high throughput that demands regularity and locality. In Table 2.1, important implementation properties for the different code classes are summarized.

Code	Dec. Algorithms	Parallel vs. Serial	Locality	Compute Kernels	Transfers vs. Compute
Turbo	MAP	Serial/iterative	Low (interleaver)	Add-Compare-Select	Compute dominated
LDPC	Belief propagation	Parallel/iterative	Low (Tanner graph)	Min-Sum/ add	Transfer dominated
Polar	Successive cancelation/ list	Serial	High	Min-Sum/add/ sorting	Balanced

Table 2.1. *Implementation properties of various coding schemes*

Functional parallelism/pipelining is an efficient technique to speed up algorithms with data dependencies. We can "unroll" the iterations and insert buffers between the different pipeline stages in iterative decoding algorithms. In this way, the various iterations can be calculated in parallel, but on different data sets. Spatial and functional parallelization are implementation techniques only, i.e. they are not changing the algorithmic behavior. We can also modify the decoding algorithm itself to enable parallelism. Let us again consider the MAP algorithm. The data dependencies in the forward/backward recursion can be broken up by exploiting the fact that the trellis has a finite memory due to the underlying constraint length of the code. This property enables the splitting of the trellis into sub-trellises that can be processed independently of each other. However, some acquisition steps are mandatory at the border of the sub-trellises to get the correct probabilities. The length of the sub-trellises and the corresponding acquisition length impact the communications performance. In this case, an increased parallelism can have a negative impact on the communications performance.

The throughput of channel decoders depends on various communication and implementation parameters. Let N be the block size and R the rate of a channel code. Let I denote the number of iterations that a corresponding (iterative) decoder requires to decode a code block ($I = 1$ in the case of a non-iterative decoding algorithm). Let P denote the degree of achievable parallelism. P is defined as the ratio between the operations that are performed in one clock cycle, and the total number of operations necessary to perform one decoding iteration for a complete code block. Note that operation represents the "critical operation" in an algorithm: an operation can be a computation as well as a data transfer. Let f be the clock frequency. Using these parameters, the throughput (in information bits per second) of a FEC architecture can be roughly estimated by [2.1]

$$T_{inf} = N \cdot R \cdot \frac{1}{I} \cdot P \cdot f \cdot (1 - \omega), \qquad [2.1]$$

where ω is a normalized value between 0 and 1, that represents the timing overhead due to, for example, data distribution, interconnect, memory access conflicts, etc. The maximum clock frequency f is determined by the critical path in the compute kernels of the corresponding decoding algorithms and/or delay due to interconnect. Moreover, f is typically upper limited to 1 GHz due to power and design methodology constraints. The overhead ω increases with increasing N and P and is larger for decoding algorithms that have limited locality and are data-transfer dominated. The impact of ω on the throughput can be considered as an effective reduction of the clock frequency f or decrease in P, if additional clock cycles are mandatory, for example, due to memory conflicts, that cannot be hidden. If we are targeting 1 Tbit/s throughput with a frequency limit of 1 GHz, 1,000 bits have to be decoded in one single clock cycle that requires extreme parallelism. To achieve the highest throughput, P has to be maximized, and ω minimized. In the following sections, we will discuss the throughput maximization for the different coding schemes.

2.3.1. *Turbo decoder*

Among the three code classes, Turbo codes are most challenging with respect to high-throughput decoding. The decoding structure consists of two component decoders connected through an interleaver/deinterleaver. These component decoders work cooperatively by exchanging extrinsic information in an iterative loop. The respective bit-wise log-likelihood ratios are computed in a forward and a backward recursion on the trellis in the MAP-based component decoder. Thus, decoding is inherently serial on component decoder (MAP recursions) and on Turbo decoder level (iterative loop). The kernel operation in the MAP is the Add-Compare-Select (ACS) operation. Depending on the encoder's memory depth, several ACS operations have to be executed to process a single trellis step in the forward and backward recursion, respectively. In total, $4 \cdot K$ trellis computations and two interleavings have to be performed for one single Turbo decoding iteration. Hence, P is determined by the number of parallel trellis step computations performed by a decoding architecture. The maximum value of P for one decoding iteration, i.e. $P = 1$, can be achieved by an architecture if the forward/backward recursions of the MAP algorithm are unrolled and pipelined (functional parallelism), yielding the XMAP architecture. $P = I$ if, in addition, iterations are unrolled and pipelined. For a detailed overview and discussion of such highly parallel decoders, we refer to (Weithoffer *et al.* 2018). Figure 2.2 (left) shows the layout of a Turbo decoder exploiting all levels of parallelism ($P = I$) and an optimized interleaver such that ω is very small (Weithoffer *et al.* 2018). The decoder achieves $102\,\mathrm{Gbit/s}$ information bit throughput for an information block length of $128\,\mathrm{bit}$, on a low V_t $28\,\mathrm{nm}$ Fully Depleted Silicon on Insulator (FD-SOI) technology under worst-case PVT conditions, running with $800\,\mathrm{MHz}$ and performing four decoding iterations. Taking into account the code rate of $R = 1/3$, the corresponding coded throughput is $306\,\mathrm{Gbit/s}$. The total area is $23.61\,\mathrm{mm}^2$. Different colors represent the eight different MAP decoders originating from the four unrolled iterations (each iteration requires two MAP decoders). Although this architecture achieves an information bit throughput of more than $100\,\mathrm{Gbit/s}$ for small block sizes, it still suffers from limited flexibility in terms of block sizes and varying number of iterations, a large area and high power consumption. Hence, further research is mandatory.

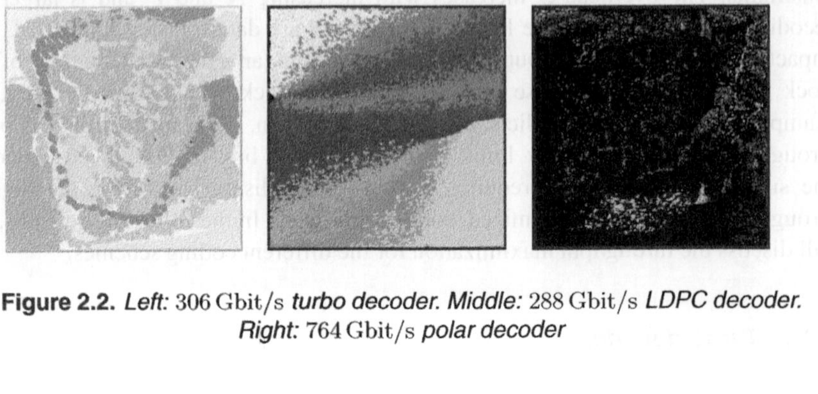

Figure 2.2. *Left:* $306\,\mathrm{Gbit/s}$ *turbo decoder. Middle:* $288\,\mathrm{Gbit/s}$ *LDPC decoder. Right:* $764\,\mathrm{Gbit/s}$ *polar decoder*

2.3.2. *LDPC decoder*

LDPC decoding is based on an iterative message exchange between variable and check nodes on the Tanner graph that is represented by the parity check matrix H. Unlike the Turbo decoding, this BP has some inherent parallelism, since all check nodes (variable nodes) can be processed independently from each other. The decoder throughput is mainly limited by the iterative data exchange between the check and variable nodes. The result of each check node or variable node calculation has to be spread via the edges of the Tanner graph to all other connected nodes. The Tanner graph has very limited locality to provide good communications performance, which challenges an efficient implementation of the data transfers and largely impacts ω. Hence, in contrast to Turbo decoding, the BP algorithm is data transfer and not compute dominated. Let us consider an LDPC block code with a parity check matrix H that has $\#edges(H)$ 1-entries (number of 1's in H equals the number of edges in the Tanner graph). Furthermore, let $\#proc\_edges(A)$ denote the number of edges that can be processed in one clock cycle by a decoder architecture. The corresponding parallelism is $P = \frac{\#proc\_edges(A)}{\#edges(H)}$. We can classify different decoder architectures. *Partially parallel architectures:* Only a subset of edges and nodes are processed in parallel, i.e. $P < 1$. These architectures are very common for large block sizes that use quasi-cyclic (QC) block codes. An example is the DVB-S2 standard. *Fully parallel architectures at iteration level:* All edges are processed in parallel, i.e. $P = 1$. All check nodes and variable nodes are instantiated as hardware units and the corresponding edges are hardwired. Because of the low locality in the Tanner graph, routing congestion (minimizing ω) is a big challenge in these architectures. *Unrolled fully parallel architectures:* These architectures are similar to the fully parallel architectures, but, in addition, the iterations are unrolled and pipelined, i.e. $P = I$. In every clock cycle, a new block is processed. Only this architectural approach is feasible to achieve a throughput towards 1 Tbit/s (Schläfer *et al.* 2013; Ghanaatian *et al.* 2018). Figure 2.2 (middle) shows the layout of an (672,420) LDPC decoder using the unrolled fully parallel architecture approach. The decoder is implemented in the same technology and under same PVT assumptions as the Turbo decoder and runs with 400 MHz, achieving a (coded) throughput of 288 Gbit/s. The total area is $3.31\ \text{mm}^2$. Different colors represent the individual iterations (in total 10).

2.3.3. *Polar decoder*

As discussed earlier, Polar codes have recently attracted much attention in the context of 5G. Successive Cancelation (SC), Successive Cancelation List (SCL) and BP are the most prominent decoding algorithms for Polar codes. Decoding corresponds to a breadth-first (BP) or depth-first traversal (SC, SCL) of the Polar Factor Tree (PFT) (Alamdar-Yazdi and Kschischang 2011), in which the received log-likelihood ratios from the channel are processed by the tree nodes. BP needs a large set of iterations to achieve the error correction performance of SC and is, thus, not well suited for very high throughput and low latency (Abbas *et al.* 2017). SC and SCL, on the other hand, have sequential behavior due to the mandatory depth-first tree traversal. To achieve a very high throughput, the tree traversal can be unrolled and pipelined (Giard *et al.* 2015), very similar to the iteration unrolling in Turbo code and LDPC code decoding as described above. Whenever a node is visited during the tree traversal, a corresponding pipeline stage can be instantiated. In this way, for a block length of N (= number of leaves in the PFT), the maximum number of pipeline stages is $2 * (2N - 2) + 1$. In this way, all $N * \log N$ operations are executed in parallel, i.e. $P = 1$. Obviously, the complexity of the decoding architecture is directly proportional to the size of the PFT. However, for a given code, the tree can be reduced by various transformations. For example, if a subtree represents a repetition code or a parity check code, the corresponding subtree can be replaced by a single node. Similarly, we can merge rate-0 and rate-1 nodes into its parent nodes (Sarkis *et al.* 2014) or use majority logic decoding in subtrees. The achievable tree reduction strongly depends on the underlying Polar code. SC is the best candidate to achieve throughput towards $1\,\text{Tbit/s}$. Here, we consider a (1,024, 512) Polar code that is decoded with the SC algorithm. Again we use the same technology and PVT setup as for the Turbo and LDPC decoders. The unoptimized PFT has 2,047 nodes, corresponding to 4,093 pipeline stages. Performing the aforementioned tree optimization yields a reduction to 385 stages. Implementing this tree in a fully pipelined architecture would require about $310\,\text{KB}$ pipeline memory. This huge memory requirement illustrates a further challenge for high-throughput decoder architectures. Although pipelining enables highest throughput for decoders, these architectures suffer from large latency and, more importantly, large power consumption. A huge amount of data has to be stored inside the pipelined architecture. These data sets are stored in registers that have to be driven by a clock tree. If a design contains many registers, the clock tree can become the dominant power consumer. Hence, minimizing the register load is a main optimization goal. On the code level, it can be performed by the aforementioned tree optimizations. On an architectural level, register balancing (retiming) is a very efficient technique to further reduce the pipeline stages. Advanced retiming with a frequency constraint of $700\,\text{MHz}$ results in a partially pipelined architecture with only 105 pipeline stages ($85\,\text{KB}$ memory). The power consumption of this design is $5.7\,\text{W}$, of which more than 70% is consumed in registers and the clock tree. In a further step, some of the registers can be replaced by latches that have a much lower load. Latch-based design with some further optimizations reduces the area from $3.14\,\text{mm}^2$ to $2.79\,\text{mm}^2$ and the power consumption from $5.7\,\text{W}$ to $2.7\,\text{W}$. The final (coded) throughput is $664\,\text{Gbit/s}$. Only 32% of the overall power accounts for registers/latches and the clock tree. Figure 2.2 (right) shows the layout of this Polar decoder. Each color represents a pipeline stage (105), the black color is memory. The decoder was fully automatically optimized and generated with the framework presented in Kestel *et al.* (2018b) and Lehnigk-Emden *et al.* (2019).

2.4. Conclusion

We have shown that throughputs towards 1 Tbit/s are feasible for all three code classes by appropriate "unrolling", using heavy pipelining and spatial parallelism. However, this architectural approach is limited to small block sizes and small numbers of iterations (Turbo and LDPC codes), which negatively impacts the communications performance. Moreover, although pipelining largely increases the throughput and locality, it also increases the latency. All architectures suffer from limited flexibility in terms of block sizes (all three codes), varying number of iterations (Turbo and LDPC codes) and code rate flexibility (LDPC and Polar codes). In summary, the biggest challenge for very high-throughput decoder architectures lies in the improvement of the communications performance, under the aforementioned implementation constraints and providing block size, code rate and algorithmic flexibility. As discussed in the introduction, microelectronic progress will largely contribute to an improved area efficiency but not as much to an increased performance and a reduced power density. Thus, further research is mandatory to keep pace with the increasing requirements on communication systems in terms of throughput, latency, power/energy efficiency, flexibility, cost and communications performance.

2.5. Acknowledgments

We gratefully acknowledge financial support by the EU (project-ID: 760150-EPIC).

2.6. References

Abbas, S.M., Fan, Y., Chen, J., and Tsui, C.Y. (2017). High-throughput and energy-efficient belief propagation polar code decoder. *IEEE Transactions on Very Large Scale Integration (VLSI) Systems*, 25(3), 1098–1111.

Alamdar-Yazdi, A. and Kschischang, F.R. (2011). A simplified successive-cancellation decoder for polar codes. *IEEE Communications Letters*, 15(12), 1378–1380.

Amdahl, G.M. (2013). Computer architecture and Amdahl's law. *Computer*, 46(12), 38–46.

Arikan, E. (2009). Channel polarization: A method for constructing capacity-achieving codes for symmetric binary-input memoryless channels. *IEEE Transactions on Information Theory*, 55(7), 3051–3073.

Berrou, C., Glavieux, A., and Thitimajshima, P. (1993). Near Shannon limit error-correcting coding and decoding: Turbo-codes. 1. *Proceedings of ICC '93 - IEEE International Conference on Communications*, vol. 2, 1064–1070.

EPIC (2020). Enabling Practical Wireless Tb/s Communications with Next Generation Channel Coding [Online]. Available at: https://epic-h2020.eu/results.

Fettweis, G.P. and Matus, E. (2017). Scalable 5G MPSoC architecture. *2017 51st Asilomar Conference on Signals, Systems, and Computers*, 613–618.

Gallager, R. (1962). Low-density parity-check codes. *IRE Transactions on Information Theory*, 8(1), 21–28.

Ghanaatian, R., Balatsoukas-Stimming, A., Müller, T.C., Meidlinger, M., Matz, G., Teman, A., and Burg, A. (2018). A 588-Gb/s LDPC decoder based on finite-alphabet message passing. *IEEE Transactions on Very Large Scale Integration (VLSI) Systems*, 26(2), 329–340.

Giard, P., Sarkis, G., Thibeault, C., and Gross, W.J. (2015). 237 Gbit/s unrolled hardware polar decoder. *Electronics Letters*, 51(10), 762–763.

Horowitz, M. (2014). Computing's Energy Problem: (and what we can do about it). *Keynote in 2014 IEEE International Solid-State Circuits Conference*. [Online]. Available at: http://eecs.oregonstate.edu/research/vlsi/teaching/ECE471_WIN15/mark_horowitz_ISSCC_2014.pdf

Ilnseher, T., Kienle, F., Weis, C., and Wehn, N. (2012). A 2.15GBit/s turbo code decoder for LTE advanced base station applications. *2012 7th International Symposium on Turbo Codes and Iterative Information Processing (ISTC)*, 21–25.

Kestel, C., Herrmann, M., and Wehn, N. (2018a). When channel coding hits the implementation wall. *2018 IEEE 10th International Symposium on Turbo Codes Iterative Information Processing (ISTC)*, 1–6.

Kestel, C., Weithoffer, S., and Wehn, N. (2018b). Polar code decoder exploration framework. *Advances in Radio Science*, 16, 43–50.

Kienle, F., Wehn, N., and Meyr, H. (2011). On complexity, energy- and implementation-efficiency of channel decoders. *IEEE Transactions on Communications*, 59(12), 3301–3310.

Lehnigk-Emden, T., Alles, M., Kestel, C., and Wehn, N. (2019). Polar code decoder framework. *2019 Design, Automation Test in Europe Conference Exhibition (DATE)*, 1208–1209.

MacKay, D.J.C. and Neal, R.M. (1997). Near Shannon limit performance of low density parity check codes. *Electronics Letters*, 33(6), 457–458.

Moore, G.E. (1965). Cramming more components onto integrated circuits. *Electronics*, 114–117.

Muller, S., Schreger, M., Kabutz, M., Alles, M., Kienle, F., and Wehn, N. (2009). A novel LDPC decoder for DVB-S2 IP. *2009 Design, Automation Test in Europe Conference Exhibition*, 1308–1313.

Sarkis, G., Giard, P., Vardy, A., Thibeault, C., and Gross, W.J. (2014). Fast polar decoders: Algorithm and implementation. *IEEE Journal on Selected Areas in Communications*, 32(5), 946–957.

Schläfer, P., Wehn, N., Alles, M., and Lehnigk-Emden, T. (2013). A new dimension of parallelism in ultra high throughput LDPC decoding. *SiPS 2013 Proceedings*, 153–158.

Scholl, S., Weithoffer, S., and Wehn, N. (2016). Advanced iterative channel coding schemes: When Shannon meets Moore. *2016 9th International Symposium on Turbo Codes and Iterative Information Processing (ISTC)*, IEEE, Brest, France, 406–411.

Shannon, C.E. (1948). A mathematical theory of communication. *The Bell System Technical Journal*, 27(4), 623–656.

Shockley, W. (1949). The theory of p–n junctions in semiconductors and p–n junction transistors. *The Bell System Technical Journal*, 28, 435–489.

Thul, M.J., Gilbert, F., Vogt, T., Kreiselmaier, G., and Wehn, N. (2005). A scalable system architecture for high-throughput turbo-decoders. *The Journal of VLSI Signal Processing-Systems for Signal, Image, and Video Technology*, vol. 39, 152–158.

Vogt, T. and Wehn, N. (2008). A reconfigurable ASIP for convolutional and turbo decoding in an SDR environment. *IEEE Transactions on Very Large Scale Integration (VLSI) Systems*, 16(10), 1309–1320.

Weithoffer, S., Nour, C., Wehn, N., Douillard, C., and Berrou, C. (2018). 25 years of turbo codes: From Mb/s to beyond 100 Gb/s. *International Symposium on Turbo Codes & Iterative Information Processing (ISTC)*, December, Hong Kong, China.

PART 2

Application-specific MPSoC Architectures

3

Automation for Industry 4.0 by using Secure LoRaWAN Edge Gateways

Marcello COPPOLA[1] and George KORNAROS[2]

[1] *STMicroelectronics, Grenoble, France*

[2] *ISCA Lab, Hellenic Mediterranean University, Heraklion, Greece*

Industrial IoT (IIoT) and Industry 4.0 signify the advent of connecting industrial automation devices and equipment, or "things", with cloud-based systems to harvest, process and analyze information faster than ever; to increase business value through novel customer experience and services (real-time analytics or visualization of data) and improve operational efficiency. On the other hand, the industry will need to reduce security risks more than ever before. Therefore, they need to protect all connected devices and environments on all fronts. In this increasingly connected and software-dominated ecosystem, IoT gateways are the key components. IoT gateways communicate to a multitude of sensors and automation controllers, called "edge devices", and provide the bridge between an on-premise communication network and cloud-based computing supremacy and visualization.

In an Industrial IoT ecosystem, security is the key aspect in which IoT gateway security is of prime importance since a secured gateway enables protection of the industrial production cycle. To guarantee safe and secure operations, service availability and improved user experience, it is mandatory to ensure sufficient

---

For a color version of all figures in this book, see www.iste.co.uk/andrade/multi2.zip.

Multi-Processor System-on-Chip 2 – Applications,
coordinated by Liliana ANDRADE and Frédéric ROUSSEAU.

security measures, thus avoiding chances of potential risks such as malicious threats, spoofing, man-in-the-middle (MITM) attacks, data snooping or privacy breaching.

In conjunction with many technological, environmental and economic benefits, the rapidly moving connected world also represents an array of growing attacks such as side-channel attacks, fault attacks and physical tampering. Given these risks, ensuring security, trustworthiness and robustness of IIoT becomes imperative, in which IIoT gateways play an important role. Providing complete IoT security not only requires that communication from the gateway to the cloud is secure, but also that the gateway can participate in the secure communication and management of connected edge node devices, which themselves must be secured. This chapter presents the key concepts required to implement a security architecture to enable complete IIoT security in a LoRaWAN environment, through the use of in-depth analysis of methods.

3.1. Introduction

The digital revolution blurred the boundaries between the physical and digital worlds within Industry 4.0, which becomes the place where employees, machines and products interact via an Industrial IoT (IIoT) system. An IIoT system leverages on CPS which enables different physical sensors, actuators and controllers to be locally interconnected and globally connected to cloud computing servers; thus forming complex online systems – a collection of independent interconnected computing elements. The novel Industry 4.0 manufacturing process implies a strong interaction between employees and products at all stages. Industry 4.0 plays a significant role in the strategy to take the opportunities of digitalization of all stages of production and service systems. The Fourth Industrial Revolution is becoming increasingly established as a result of a combination of numerous physical and digital technologies such as artificial intelligence, cloud computing, adaptive robotics, augmented reality, additive manufacturing and Internet of Things (IoT). As an industrial IoT platform (IIoT) is composed of IIoT devices that enable objects, assets, machinery, embed sensors and tags to be connected, and establish an Internet-enabled connection to transmit data and remotely monitor and control the manufacturing production line. Generally, an IIoT device is composed of a microcontroller (e.g. an STM32) that acts as the brains of the IIoT device itself, where the application runs, a *communication module* (such as Wi-Fi, BLE and LORA) and one or more *sensors* or *actuators* of some kind. The microcontroller handles the application, data acquisition and communication with the Cloud Service. On top of these IIoT platforms, several services can be provided by IIoT solution providers. Services are what the user sees and interacts with. Since these services use the Internet, it is important for IIoT solution providers to make sure that the IIoT device is trustworthy, because they wish to protect their services, network and brand.

As IIoT deployments have grown, so have threats to security. According to the SonicWall Cyber Threat Report (SonicWall 2019), through the first six months of 2019, 2.4 million encrypted attacks on IoT were registered, almost eclipsing the 2018 full-year total in half the time. Security has to be considered from the very beginning of the design phase so as to mitigate vulnerabilities and threats. In order to be considered trustworthy, IIoT devices need to be legitimate, identifiable and manageable in order to prevent hackers, counterfeits, ODM overproduction and IP theft. One reason for the difficulty in securing connected devices is that many stakeholders are involved in the IIoT device lifecycle, with each of them facing different threat vectors, implying different security requirements. Since OEMs, manufacturers, integrators and designers are involved in developing and implementing the IIoT devices, it becomes quite complex to legitimate and to identity them all.

The generic architecture for IIoT solutions includes IIoT devices that are connected to the private – or a private area – of the public cloud (e.g. AWS IoT Core) using some sort of network. Thus, IIoT devices can be linked to services that are executed in the Cloud, related to analytics, billing, CRM, NoC, etc. (e.g. AWS services, Chargify, Tableau, Saleforce) and other devices. Generally, Cloud providers try to secure data and interactions, process the data transmitted by different devices and trigger actions accordingly. In addition, they allow IIoT applications to interact with IIoT devices or machines, even when those devices/machines are not connected.

As it stands today, the IIoT industry does not currently have a widely used and accepted standard, provided by vendors, on how IIoT devices should connect to each other. This makes it difficult for developers to understand how this can be achieved across multiple manufacturers when developing IIoT applications. The traditional way to connect IoT devices to the Cloud is by specific SDKs. For instance, the SDK for AWS IoT devices makes it possible to connect and authenticate IoT devices to AWS IoT Core, as well as exchange messages with the service, via MQTT, HTTP or WebSockets protocols. This SDK for AWS IoT devices supports C, JavaScript and Arduino and includes client libraries, a manual for developers and a porting guide for manufacturers. Although these solutions address security, they are not able to mitigate vulnerabilities and threats. As a matter of fact, in Industry 4.0, hence IIoT security needs to combine IIoT device security with network security. The IIoT device security requires ensuring a trusted and authentic software and firmware execution that implies a secure and verified boot; while the network security needs to ensure data integrity, authentication via unique device identity and data communication protection.

The most important decision when designing an IIoT device is the choice of a robust root-of-trust management. This is a basic need and is today the weakest link in the IoT. Proving ownership of the root of trust is the baseline for cryptographic and security foundations to enable safe data communication and functional operation of IIoT networks. It is widely acknowledged in the security industry that strong security mechanisms have to be based on hardware because software can always be circumvented by software. Any IIoT solution has to combine operational technology (OT) and information technology (IT) in order to implement **machine to machine connectivity, to better monitor the manufacturing process and to control the physical machine remotely**. In IIoT the IT network plays an important role since it moves data from the sensor to the backhaul that takes it up to the Internet. Several low-power wide-area network (LPWAN) (Raza *et al.* 2017) technologies have been proposed as an alternative solution for implementing private cellular-like networks, with infrastructure costs as low as possible. Coverage, scalability and energy efficiency requirements of Internet of Things (IoT) deployments have recently steered the attention of the research community, industry and several network operators to long-range wide-area networks (LoRaWAN) (Augustin *et al.* 2016; Palattella *et al.* 2016; Raza *et al.* 2017). It integrates the long-range (LoRa) technology at the physical (PHY) layer, provides specifications for both media access control (MAC) and network layers and defines the four key components of the overall architecture, which are end nodes, gateways, cloud/network servers and remote applications (LoRa Alliance Technical Committee 2017).

OT is responsible for monitoring and managing industrial process assets and manufacturing equipment and is a technology that provides a way to interface the physical world. OT includes industrial control systems (ICS), supervisory control and data acquisition (SCADA) and distributed control systems (DCS). As defined by Gartner, it is: "Hardware and software that detects or causes a change through the direct monitoring and/or control of physical devices, processes and events in the enterprise". Combining OT with IT-based technologies such as big data and the Internet of Things, we move to IIoT that opens the door to novel services within the industrial processes, such as remote diagnostics and predictive maintenance. As a result, physical machines will produce a large amount of data, which, if properly analyzed, can improve the management of the manufacturing process. Since connectivity opens a door to the outside, it is important to tackle all the malware, identity management and access control security challenges that may arise for OT, since unmanaged vulnerabilities in an OT system can leave critical infrastructure at risk of sabotage. Thus, the ongoing digitization required security. There are also important differences between IT and OT security. The latter, traditionally, was more about protecting physical processes, safety, uptime/production/efficiency and protection of people, while IT security is more oriented towards protecting all aspects of data and how information is stored, transmitted, processed and used in business processes.

3.2. Security in IIoT

Given the trends and tremendous growth of IoT environments, privacy and security have been of prime importance in both industrial and commercial requirements of IoT devices. The security threats and privacy concerns in each layer of the IoT architecture have now become an exigent area of research (Abomhara and Køien 2015; Chen *et al.* 2018; Sisinni *et al.* 2018; Bhat and Dutta 2019; Hassija *et al.* 2019). The primary principles to guarantee security in the entire IIoT system involve the following:

– **Confidentiality**: Confidentiality is the ability to hide information from people who are unauthorized to access it and thus needs protection from unauthorized access. Confidentiality is an important security feature in IoT. In most situations and scenarios, sensitive data, for instance, patient data, private trade data and/or military data, as well as security credentials and secret keys, must be hidden from unauthorized accesses.

– **Integrity**: Integrity of information refers to protecting information from unauthorized, unanticipated or unintentional modification. Integrity is a mandatory security property in most cases in order to provide reliable services to IoT entities (i.e. devices, users). Different systems in IoT have diverse integrity needs.

– **Availability**: Availability is the access to information whenever needed by a user of a device (or the device itself). Therefore, IoT resources must be available on a timely basis to meet needs or to avoid significant losses.

– **Authenticity**: The authenticity property allows identification of the sender entity and, further, only authorized entities to perform certain operations in the device or network. Different authentication needs require different solutions. Some solutions must be strong control, for example, authentication of finance systems.

– **Non-repudiation**: IoT services must provide a trusted audit trail. The property of non-repudiation presents certain evidence in cases where the user or device cannot deny an action, for instance, payment action.

– **Privacy**: Privacy is an entity's right to determine the degree to which it will interact with its environment and to what extent the entity is willing to share personal information with others.

3.3. LoRaWAN security in IIoT

To address wireless security concerns, LoRaWAN uses two layers of security, as can be seen in Figure 3.1: one for the network and one for the application layer. Network layer security ensures the authenticity of the device in the network. Application layer security ensures that the network operator does not have access to the end user's application data.

An end device (node) is activated before it can communicate on the LoRaWAN network, either by using *over-the-air activation* (OTAA) or the *activation by personalization* (ABP) method. OTAA is the recommended activation method for higher-security applications (LoRa Alliance Technical Committee 2017).

The first method uses over-the-air *join requests* coupled with *join accept* messages. Each end device (node) is deployed with a 64-bit DevEUI, a 64-bit AppEUI and a

128-bit AppKey. The DevEUI is a globally unique identifier for the device which has a 64-bit address, comparable with the MAC address for a TCP/IP device. The AppKey is used to cryptographically sign the join request. All three values are then made available to the application server to which the device is supposed to connect. The AppKey is used when the node sends a join request message. The node sends the join request message, composed of its AppEUI and DevEUI. It additionally sends a DevNonce, which is a unique, randomly generated, 2-byte value used for preventing replay attacks.

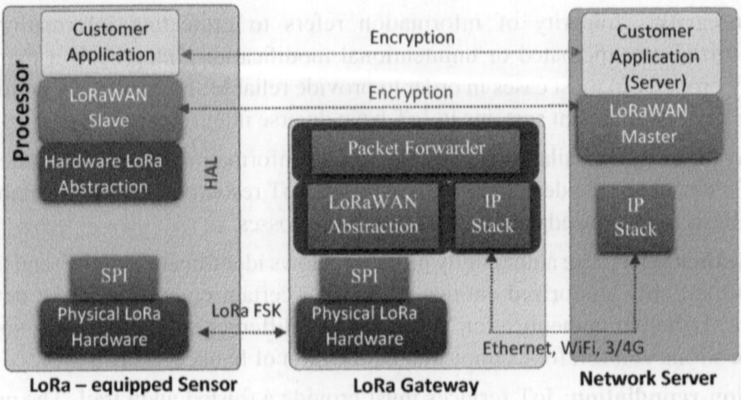

Figure 3.1. *Security in LoRaWAN*

These three values are signed with a 4-byte MIC (message integrity code) using the device's AppKey. The server accepts join requests only from devices with known DevEUI and AppEUI values while validating the MIC using the AppKey. If the server accepts the join request, it responds to the device with a join accept message. The application and network servers calculate the node's two 128-bit keys: the application session key (AppSKey) and the network session key (NwkSKey). These are calculated based on the values sent in the join request message from the node.

Additionally, the application server generates its own nonce value: AppNonce. This is another unique, randomly generated value. The join accept reply includes the AppNonce, a NetID and an end device address (DevAddr) along with configuration data for RF delays (RxDelay) and channels to use (CFList). The device address (DevAddr) in the join accept reply is a 32-bit identifier, which is unique within the network. It is possible to use the device address to differentiate between end devices which have already joined the network. This allows the network and application servers to use correct encryption keys and to properly interpret the data. When receiving data back, the data is encrypted with the AppKey. The node then uses the

AppKey to decrypt the data and derives the AppSKey and the NwkSKey using the AppNonce value received in the join accept reply.

3.4. Threat model

The threat model we consider involves attacks seeking to perform malicious network or firmware attacks, by sniffing and replaying network traffic, by modifying the IoT node firmware or by injecting malicious code into the firmware, the operating system or the kernel driver. Because the presence of a malicious manufacturer is not a consideration for our attack model, intentionally implanted vulnerabilities or malwares are not assumed to be loaded.

3.4.1. *LoRaWAN attack model*

In the OTAA method, details are exchanged (using join request and join accept messages) and security contexts are generated. The generated security contexts make sure that both the end device and the network are using the correct security keys for encryption and message signing. The join accept message however is vulnerable to an eavesdropping and replay attack combination. The first step in OTAA is that the end device transmits a join request message to join the network. Such a message contains an AppEUI, DevEUI and DevNonce. The AppEUI and DevEUI are unique addresses in the EUI-64 address space. The DevNonce is a random value which is used to prevent replay attacks. The join request message is processed by the network server. If the network server accepts the end device, a join accept message is generated and returned to the end device. The join accept message contains details (such as AppNonce and NetID) that are used for generating a correct security context (NwkSKey, AppSKey, frame counters, etc.) so that the end device uses the same security keys as the network. Unfortunately, the join accept message does not include a reference to the join request message that it was generated for, and therefore OTAA is vulnerable to replaying older join accept messages addressed to the same end device. Additionally, the network server tracks not all but only the most recently used N DevNonce values. These vulnerabilities in the LoRaWAN specification v1.0.2 have been successfully mitigated in v1.1; first by using an incrementing JoinNonce value to prevent replay attacks of Join-Accept messages, and second, a v1.1 end device does not reuse a DevNonce value (LoRa Alliance Technical Committee 2017). However, LoRaWAN specification v1.1 includes a compatibility mode for v1.0.2-only end devices. For these devices, this vulnerability is still present, even though the network is running on v1.1. Further, a malicious network server is of major concern, since by replaying OTA activation messages to the join server, this may in turn cause the overflow of a counter-nonce, and the compromise of the integrity and confidentiality of application data.

Root keys are used for protecting the integrity of join request messages and protecting the confidentiality of join accept messages, triggered by join request messages. These root keys stored in the end devices and in the join server are essential to the overall security of LoRaWAN. Hence, these keys must be protected against physical tampering. The specification leaves secure provisioning, storage and usage of root keys out of scope, but points to SE (secure elements) and HSM (hardware security modules) as possible enablers for a solution. LoRaWAN does not feature OTA firmware updates. This makes issuing fixes for the discovered vulnerabilities to the deployed end devices impractical and also makes it more likely that different versions of the protocol will coexist in a network. This, in turn, elevates the importance of addressing security issues related to backward compatibility.

Finally, it has been demonstrated that the LoRa transmissions are prone to jamming attacks (Reynders *et al.* 2016). Specifically, simultaneous communications that use the same spreading factor and frequency can conflict with each other. Although it is hard to prevent jamming attacks, by detecting such abnormal behavior, network administrators can take appropriate action (e.g. by switching the operational frequency) to prevent the impact of jamming.

3.4.2. *IIoT node attack model*

An attacker can perform malicious firmware attacks by modifying the firmware of an IoT node or injecting malicious code into the firmware. Attackers with physical access to the device can even perform cold-boot attacks to recover (even imperfect) keys from on-device persistent memory (flash) or dynamically compromise the firmware running in DRAM memory. Attacks come in a large variety of forms and may include side-channel attacks, where the physical realization of a cryptographic primitive can leak additional information, such as computation time, power consumption and radiation/noise/heat emission.

During firmware execution, an adversary can seek to exploit the dynamic memory regions (stack and heap) to circumvent secure boot and attestation. Exposure at execution time can also come from firmware's vulnerabilities such as code injection and code reuse attacks through buffer overflow (when new data exceed predefined fixed size) (Bishop *et al.* 2012) and privilege escalation, or API abuse. In a form of code reuse attack, called return-oriented programming (Buchanan *et al.* 2008) or a return-flow hijacking attack, the address of different code snippets is put on the stack (by leveraging a stack overwrite vulnerability) and it then exploits the return address of a function to jump to this code snippet.

We do not consider cache/timing side-channel attacks (Lipp *et al.* 2016) and hardware attacks, for example, cold-boot (Halderman *et al.* 2009), bus snooping (EPN Solutions 2014) and rowhammer (van der Veen *et al.* 2016). Defense techniques against these attacks can be adopted in the future.

3.5. Trusted boot chain with STM32MP1

The primary principle to establish the foundation of trust in the service and data of a device is based on the auditable and securely signed history of operations of a software running on the device.

3.5.1. *Trust base of node*

Modern commercial embedded devices commonly comprise ARM-based SoC hardware with security extensions. These nodes encompass ARM TrustZone to play a role in keeping security-sensitive resources safe and to execute security-critical services. We assume that commercially available security features such as secure boot (ARM Ltd. 2009) and active monitoring, are already activated as a base-line defense against attacks. Thus, a trusted execution environment (TEE) that is isolated from a rich execution environment (REE) aims to protect assets such as crypto keys and user credentials.

In such a system, we consider attackers who can use any existing system vulnerability to hack into the normal-world or secure-world operating system, either by loading new unverified code binaries or by modifying code binaries that were previously mapped and already exist in the memory. Thus, we consider attacks that transiently manipulate the legitimate code in user mode or map the malicious code in the data region of a pre-authorized application, but additionally attempt to leak secure critical data by malicious code or untrusted operating systems.

3.5.2. *Trusted firmware in STM32MP1*

Trusted firmware-A (TF-A) is a reference implementation of secure-world software for Arm A-Profile architectures (Armv8-A and Armv7-A), including an exception level 3 (EL3) secure monitor (ST 2019; TrustedFirmware Association 2019). It provides a suitable starting point for productization of secure world boot and runtime firmware, in either the AArch32 or AArch64 execution states. Although it was first designed for Armv8-A platforms, it has been adapted to be used on Armv7-A platforms by STMicroelectronics.

To boot Linux in an STM32MP1 device, several steps are required to progressively initialize the platform peripherals and memories. In an STM32MP1 device, TF-A is a reduced version of the Linux kernel, which is configured with only the devices used during boot via the device tree. TF-A consists of the following binaries: BL1, BL2 and BL32, where the full boot sequence in AArch32 (ARM 32-bit processors) defines:

– boot loader stage 1 (BL1) application processor-trusted ROM;

– boot loader stage 2 (BL2) trusted boot firmware;

– boot loader stage 3-2 (BL32) runtime software;

– boot loader stage 3-3 (BL33) non-trusted firmware.

For STM32 MPU platforms, BL33 is the secondary stage boot loader (SSBL), which is U-Boot by default and is the first non-secure code loaded by TF-A. Figure 3.2 depicts the trusted boot chain process in an STM32MP1 device.

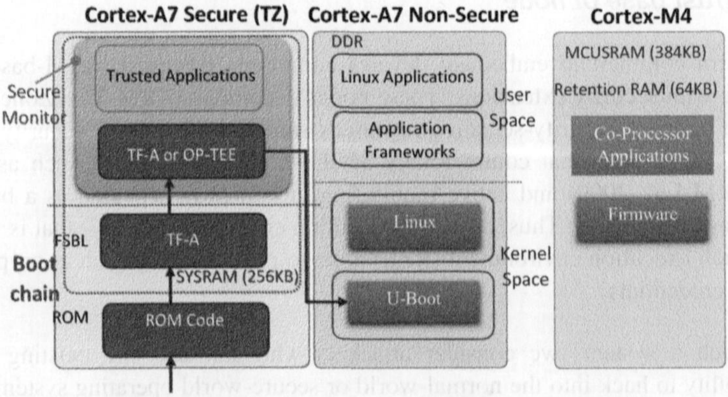

Figure 3.2. *Boot process in an STM32MP1 device*

By using asymmetric cryptography, the boot process is as follows: software image hash is encrypted with the private key and is deployed with its signature on device flash. The public key is burned inside STM32MP1 chip's OTP zone. After enabling the secure boot feature, at every boot, the STM32MP1 device extracts the image signature, decrypts it with the public key, calculates the software image hash and compares the two hashes: software will be executed only if they are equal.

3.5.3. *Trusted execution environments and OP-TEE*

OP-TEE (Open Portable TEE, an open-source ARM TrustZone® solution) is a *trusted execution environment* (TEE) designed as a companion to a non-secure Linux kernel running on Arm Cortex®-A cores using the Arm TrustZone technology. TrustZone has been used as a cornerstone hardware technology for enabling TEE on ARM-based platforms which share the majority of mobile and embedded markets. OP-TEE implements TEE Internal Core API v1.1.x, which is the API that is exposed to trusted applications, and the TEE Client API v1.0, which is the API that describes how to communicate with a TEE. Those APIs are defined in the GlobalPlatform API specifications. OP-TEE is designed primarily to rely on the ARM TrustZone technology as the underlying hardware isolation mechanism, to enable a trusted execution environment, as shown in Figure 3.3. However, it has been structured to be compatible with any isolation technology suitable for the TEE concept and goals, such as running as a virtual machine or on a dedicated CPU.

A *rich execution environment* (REE) is an execution environment that involves at least one device and all its components or an OS, excluding any trusted or secure component. In contrast, a trusted execution environment (TEE) provides a level of security to protect against attacks and secures data access. The TEE executes alongside the REE, but is shielded from it. A trusted application executes inside a TEE and exposes secure services to applications in the REE. To exploit the trusted functionalities provided by TEE, applications need to be split into client application (CA) and trusted application (TA), where the latter holds the security-sensitive functions (i.e. implementation of encryption algorithm).

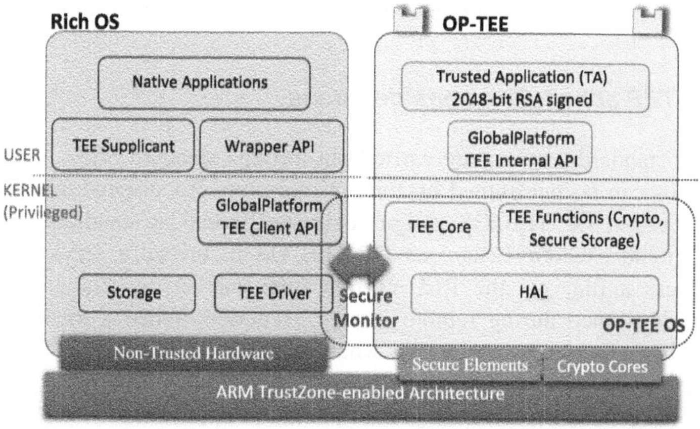

Figure 3.3. *Execution environments in OP-TEE enabled organization based on ARM TrustZone architecture*

ARM processors with TrustZone implement architectural security extensions in which each of the physical processor cores provides two virtual cores, or security domains, one being considered non-secure and called *non-secure world*, the other being considered secure and called *secure world*, and a mechanism to context switch between the two, known as the *monitor mode*.

TrustZone is based on the principle of *least privilege*, which means system modules, such as drivers and applications, do not have access to a resource unless necessary. The concept of secure (trusted) and non-secure (non-trusted) worlds extends beyond the processor to encompass memory, software, bus transactions, interrupts and peripherals within an SoC. By creating a security subsystem, assets can be protected from software attacks and common hardware attacks. TrustZone can secure a software application, library or an entire OS to run in the secure area. Non-secure software is blocked from access to the secure side and resources that reside there.

When secure mode is active, the software running on the CPU has a different view on the whole system than software running in non-secure mode. Accordingly, system functions and, in particular, security functions and cryptographic credentials can be hidden from the normal world.

OP-TEE OS executes in the secure world. World switch is done by the core's secure monitor level/mode, referred next as the monitor. Switches of world execution context are handled based on SMC exceptions and interrupt notifications. Interrupt notifications are IRQ/FIQ exceptions which may also imply switching of world execution context: normal world to secure world, or secure world to normal world.

3.5.4. *OP-TEE scheduling considerations*

OP-TEE standard services are carried out through standard SMC. Execution of these services can be interrupted by foreign interrupts. To suspend and restore the service execution, OP-TEE OS assigns a trusted thread at standard SMC entry. A trusted thread execution can lead OP-TEE OS to invoke a service in normal world: access a file, get the REE current time and so on. The trusted thread is suspended/resumed during remote service execution. When a trusted thread is interrupted by a foreign interrupt and when OP-TEE OS invokes a normal world service, the normal world gets the opportunity to reschedule the running applications. The trusted thread will resume only once the client application is scheduled back. Thus, a trusted thread execution follows the scheduling of the normal world caller context. OP-TEE OS does not implement any thread scheduling. Each trusted thread is expected to track a service that is invoked from the normal world and should return to it with an execution status.

3.5.5. *OP-TEE memory management*

OP-TEE currently requires more than 256 KB RAM for OP-TEE kernel memory. This is not a problem if OP-TEE uses TrustZone-protected DDR, but for security reasons, OP-TEE can be configured to use TrustZone-protected SRAM instead. TrustZone-protected SRAM is generally considered more secure than TrustZone-protected DRAM as there are usually more attack vectors on DRAM. The amount of available SRAM varies between platforms, from just a few KiB, up to over 512 KiB. Platforms with just a few KiB of SRAM cannot be expected to be able to run a complete TEE solution in SRAM. But those with 128 to 256 KiB of SRAM can be expected to have an appropriate TEE solution in SRAM. The *pager* provides a solution to this by demand paging parts of OP-TEE using virtual memory.

To prevent physical attacks, *pager* is a demand paging system separated from the OP-TEE kernel, which is responsible for maintaining execution of the rest components of OP-TEE kernel and TAs. It sets the on-chip SRAM memory (OCM) as the working memory for CPU to execute the OP-TEE system, and uses the DRAM as a backing store. *Pager* runs on the OCM, and the other components of the OP-TEE kernel and TAs are encrypted and stored in the DRAM to guarantee the confidentiality and integrity properties for the backing store. *Pager* manages swapping between the OCM and DRAM: when code or data stored in DRAM is demanded, *pager* decrypts and performs integrity checks on the corresponding page and loads it into the OCM; when an OCM page needs to be swapped to DRAM, *pager* encrypts it.

OP-TEE uses several L1 translation tables, one large spanning 4 GiB and two or more small tables spanning 32 MiB. The large translation table handles kernel mode mapping and matches all addresses not covered by the small translation tables. The small translation tables are assigned per thread and cover the mapping of the virtual memory space for one TA context.

Shared memory is a block of memory that is shared between the non-secure world and the secure world. It is used to transfer data between both worlds. The shared memory is allocated and managed by the non-secure world, i.e. the Linux OP-TEE driver. The secure world only considers the individual shared buffers, not their pool.

The Linux kernel driver for OP-TEE is responsible for allocating chunks of shared memory. OP-TEE Linux kernel driver relies on Linux kernel generic allocation support (CONFIG_GENERIC_ALLOCATION) to allocation/release of shared memory physical chunks. OP-TEE Linux kernel driver relies on Linux kernel dma-buf support (CONFIG_DMA_SHARED_BUFFER) to track shared memory buffer references.

3.5.6. *OP-TEE client API*

The TEE Client API describes and defines how a client running in a rich operating environment (REE) should communicate with the TEE. To identify a trusted application (TA) to be used, the client provides a UUID. All TAs expose one or several functions, which correspond to a so-called commandID, which is also sent by the client.

The TEE context is used for creating a logical connection between the client and the TEE. The context must be initialized before the TEE session can be created. When the client has completed a job running in secure world, it should finalize the context and thereby also release resources.

Sessions are used to create logical connections between a client and a specific trusted application. When the session has been established, the client has opened up the communication channel towards the specified trusted application identified by the UUID. At this stage, the client and the trusted application can start to exchange data.

3.5.7. *TEE internal core API*

The internal core API is the API that is exposed to the trusted applications running in the secure world. The TEE internal API consists of four major parts:

1. trusted storage API for data and keys;
2. cryptographic operations API;
3. time API;
4. arithmetical API.

All trusted applications (TAs) are signed with the pre-generated 2048-bit RSA development key (private key). This key is located in the keys folder (in the root of `optee_os.git`) and is named `default_ta.pem`. This key must be replaced with your own key and you should never ever check-in this private key in the source code tree when in use in a real product. The recommended way to store private keys is to use some kind of HSM (hardware security module), but an alternative would be to temporarily put the private key on a computer considered as secure when you are about to sign TAs intended to be used in real products.

3.5.8. *Root and chain of trust*

To be able to assure that devices are running, the (untampered) binaries were originally intended to run some kind of trust anchor on the devices needs to be established. The most common way of doing that is to put the root public key in some read-only memory on the device. Quite often, SoC/OEM stores public key(s) directly or the hash(es) of the public key(s) in one-time-programmable memory (OTP). When the boot ROM (which indeed needs to be ROM) is about to load the first-stage bootloader, it typically reads the public key from the software binary itself, hashes the key and compares it to the key in OTP. If they are matching, then the boot ROM can be sure that the first-stage bootloader was indeed signed with the corresponding private key.

3.5.9. *Hardware unique key*

Most devices have some kind of hardware unique key (HUK) that is mainly used to derive other keys. The HUK could, for example, be used when deriving keys used in secure storage and so on. The important thing with the HUK is that it needs to be well protected, and in the best case, the HUK should never be readable directly from software, not even from the secure side. There are different solutions to this, crypto accelerator might have support for it, or it could involve another secure co-processor.

In OP-TEE, the HUK is just stubbed and you will see that in the function called `tee_otp_get_hw_unique_key()` in `core/include/kernel/tee_common_otp.h`. In a real secure product, you must replace this with something else. If your device lacks the hardware support for a HUK, then you must at least change this to something else than just zeroes. But remember it is not a good secure practice to store a key in software, especially not the key that is the root for everything else, so this is not something we recommend that you should do.

3.5.10. *Secure clock*

The Time API in the GlobalPlatform Internal Core API specification defines three sources of time: system time, TA persistent time and REE time. The REE time is by nature considered as an unsecure source of time, but the other two should, in a fully trustable hardware, make use of a trustable source of time, i.e., a secure clock. Note that from the GlobalPlatform point of view it is not required to make use of a secure clock, i.e. it is acceptable to use time from REE, but the level of trust should be reflected by the *gpd.tee.systemTime.protectionLevel* property and the *gpd.tee.TAPersistentTime.protectionLevel* property (100=REE controlled clock, 1000=TEE controlled clock). Therefore, the functions that we need to pay attention to are `tee_time_get_sys_time()` and `tee_time_get_ta_time()`. If the hardware has a secure clock, then this implementation may change there to instead use the secure clock (and then update the property accordingly, i.e. `tee_time_get_sys_time_protection_level()` and the variable `ta_time_prot_lvl` in `tee_svc.c`).

3.5.11. *Cryptographic operations*

By default, OP-TEE uses a software crypto library (currently mbed TLS and LibTomCrypt), while there is the ability to enable crypto extensions that were introduced with ARMv8-A (if the device is capable of that). By default, OP-TEE is configured with a software PRNG. The entropy is added to software PRNG at various places, but unfortunately, it is still quite easy to predict the data added as entropy. As a result, unless the RNG is based on hardware, the generated random will be quite weak.

In OP-TEE, TAs can use services accessible through the GlobalPlatform Internal Core API implemented in *libutee*. TAs are statically linked against *libutee*, which wraps the API functions around assembler macros to OP-TEE OS system calls. The library provides interfaces to secure storage, time, arithmetic and cryptographic operations.

The secure storage API encrypts data objects by the use of a secure storage service. The encryption process involves three keys: secure storage key (SSK), trusted application storage key (TSK) and file encryption key. The SSK is generated from the hardware unique key and is used to derive TSKs. Each TA has a TSK that is generated from the SSK and the TAs UUID. Both SSK and TSK are generated using the HMAC SHA256 algorithm. Finally, for every file created, an FEK is generated from the pseudo-random number generator. The encrypted data objects are then transferred to the tee-supplicant by a series of remote procedure calls, and stored in a special file. OP-TEE further provides TAs with libraries for TLS and SSL protocols (*libmbedtls*) (ARM Ltd. 2019), arithmetic (*libmpa*) and a subset of ISO C functions (*libutils*). Once the REE application has no further service requests, the session is terminated and the context is destroyed.

3.6. LoRaWAN gateway with STM32MP1

The LoRaWAN gateway is realized through integrating the RAK831 RF front-end LoRaWAN gateway concentrator module (Shenzhen RAKwireless Technology Co., Ltd 2020) with the host system STM32MP1-DK2 connected via SPI. The LoRaWAN gateway, shown in Figure 3.4, connects to the *The Things Network* (TTN) where it publishes temperature and humidity, which are sent by a LoRa end node.

Figure 3.4. *LoraWAN gateway using an RAK831 RF with a GPS (top two shields), the STM32MP1 and a 3G modem (bottom shield) when a wired connection is not available*

The gateway packet forwarder executes inside the STM32MP1 platform, which supports the full trusted boot chain and OP-TEE. The packet forwarder has been extended by using OP-TEE infrastructure for secure storage with the *gwsecstore* library, as depicted in Figure 3.5. We developed this library consistent with the specifications of GlobalPlatform's TEE Internal Core API, to enable the packet forwarder to secure the gateway identifier. The gateway packet forwarder requires a *gateway ID*, as configured by the TTN network to initialize and connect to it. This *gateway ID* is now provided by the trusted application part of the packet forwarder. The API calls of the *gwsecstore* library follow next.

```
- TEEC_Result read_secure_object(char *id, char *data, size_t data_len);
- TEEC_Result write_secure_object(char *id, char *data, size_t data_len);
- TEEC_Result delete_secure_object(char *id);
```

Initially, in the packet forwarder setup phase, the *gateway ID*, which is retrieved from the TTN, is stored securely, while the created object ID is stored to the file system. Thus, to connect to the TTN, the packet forwarder retrieves the *gateway ID* through the *gwsecstore* library by providing the known object ID and stored object data length.

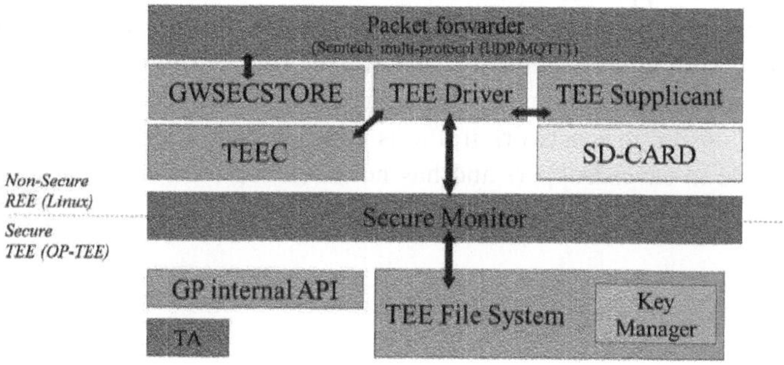

Figure 3.5. *Execution of gateway packet forwarder in OP-TEE enabled organization based on the STM32MP1 platform*

To implement secure storage in OP-TEE, it is implemented according to what has been defined in GlobalPlatform's trusted storage specification, which must provide a minimum level of protection against rollback attacks. Currently, the REE filesystem is used while in the near future we will make use of the replay protected memory block (RPMB)-based secure storage.

3.7. Discussion and future scope

We anticipate several counter-measures against different attacks. Through exploiting the trusted boot chain and OP-TEE, we address the security operation of the LoRaWAN gateway according to the specified threat model. However, consider that ARM's TrustZone technology is designed mainly to resist software attacks and it is difficult for it to resist physical attacks (Carru 2017). For systems that use sensitive data for a short amount of time (e.g. cryptographic keys), zeroing out the data immediately after use (Halderman *et al.* 2009) would significantly reduce the risk from Rowhammer-like attacks. While this counter-measure is effective for protecting short-lived data, it cannot be used for data that reside in memory for a long duration.

In terms of rapid deployment of LoRaWAN since 2018, we are working towards developing features such as firmware updating over the air. However, lack of OTA firmware updates is challenging for full migration to LoRaWAN v1.1, even if all the deployed end devices are capable of running version 1.1. Potential security problems caused by backward compatibility of LoRaWAN are twofold: a fallback may enable the exploitation of vulnerabilities associated with the older version, or it may break a feature introduced in the new version.

Even though LoRaWAN security design adheres to state-of-the-art principles, use of standards, NIST-approved algorithms and end-to-end security, secure devices of low complexity and cost must also be future-proof. The full firmware stack executing in these devices should be authentic and trustworthy during all its lifetime. This guarantee, combined with mutual device authentication and servers' security counter-measures, can ensure that network traffic is coming from a legitimate device, is not comprehensible to eavesdroppers and has not been captured and replayed by rogue actors.

3.8. Acknowledgments

The research leading to these results received funding from the European Union (EU) Horizon 2020 project AVANGARD (advanced manufacturing solutions tightly aligned with business needs) under grant agreement no. 869986.

3.9. References

Abomhara, M. and Køien, G.M. (2015). Cyber security and the internet of things: Vulnerabilities, threats, intruders and attacks. *Journal of Cyber Security*, 4, 65–88.

ARM Ltd. (2009). ARM Security Technology Building a Secure System using TrustZone® Technology [Online]. Available: http://infocenter.arm.com/help/index.jsp?topic=/com.arm.doc.prd29-genc-009492c/index.html.

ARM Ltd. (2019). Mbed TLS [Online]. Available: https://tls.mbed.org/.

Augustin, A., Yi, J., Clausen, T.H., and Townsley, W. (2016). A study of LoRa: Long range & low power networks for the Internet of things. *Sensors*, 16, 1466.

Bhat, P. and Dutta, K. (2019). A survey on various threats and current state of security in android platform. *ACM Computing Surveys*, 52, 1–35.

Bishop, M., Engle, S., Howard, D., and Whalen, S. (2012). A taxonomy of buffer overflow characteristics. *IEEE Transactions on Dependable and Secure Computing - TDSC*, 9(3), 305–317.

Buchanan, E., Roemer, R., Shacham, H., and Savage, S. (2008). When good instructions go bad: Generalizing return-oriented programming to RISC. *Proceedings of the 15th ACM Conference on Computer and Communications Security*, pp. 27–38.

Carru, P. (2017). Attack TrustZone with Rowhammer, GreHack [Online]. Available: https://grehack.fr/data/2017/slides/GreHack17_Attack_TrustZone_with_Rowhammer.pdf.

Chen, K., Zhang, S., Li, Z., Zhang, Y., Deng, Q., Ray, S., and Jin, Y. (2018). Internet-of-things security and vulnerabilities: Taxonomy, challenges, and practice. *Journal of Hardware and Systems Security*, 2(2), 97–110.

EPN Solutions. (2014). Analysis tools for DDR1, DDR2, DDR3, embedded DDR and Fully Buffered DIMM modules [Online]. Available: http://www.epnsolutions.net/ddr.html.

Halderman, J., Schoen, S., Heninger, N., Clarkson, W., Paul, W., Calandrino, J., Feldman, A., Appelbaum, J., and Felten, E. (2009). Lest we remember: Cold-boot attacks on encryption keys. *Communications of the ACM*, 52(5), 91–98.

Hassija, V., Chamola, V., Saxena, V., Jain, D., Goyal, P., and Sikdar, B. (2019). A survey on IoT security: Application areas, security threats, and solution architectures. *IEEE Access*, 7, 82721–82743.

Lipp, M., Gruss, D., Spreitzer, R., Maurice, C., and Mangard, S. (2016). ARMageddon: Cache attacks on mobile devices. *Proceedings of the 25th USENIX Conference on Security Symposium*, pp. 549–564.

LoRa Alliance Technical Committee (2017). LoRaWAN™ 1.1 Specification [Online]. Available: https://lora-alliance.org/resource-hub/lorawanr-specification-v11.

Palattella, M.R., Dohler, M., Grieco, A., Rizzo, G., Torsner, J., Engel, T., and Ladid, L. (2016). Internet of things in the 5G Era: Enablers, architecture, and business models. *IEEE Journal on Selected Areas in Communications*, 34(3), 510–527.

Raza, U., Kulkarni, P., and Sooriyabandara, M. (2017). Low power wide area networks: An overview. *IEEE Communications Surveys Tutorials*, 19(2), 855–873.

Reynders, B., Meert, W., and Pollin, S. (2016). Range and coexistence analysis of long range unlicensed communication. *Proceedings of the 23rd International Conference on Telecommunications (ICT)*, pp. 1–6.

Shenzhen RAKwireless Technology Co., Ltd. (2020). RAK831 LoRaWAN® Gateway Concentrator [Online]. Available: https://doc.rakwireless.com/datasheet/rakproducts/-.

Sisinni, E., Saifullah, A., Han, S., Jennehag, U., and Gidlund, M. (2018). Industrial Internet of things: Challenges, opportunities, and directions. *IEEE Transactions on Industrial Informatics*, 14(11), 4724–4734.

SonicWall (2019). SonicWall Cyber Threat Report [Online]. Available: https://www.sonicwall.com/resources/white-papers/2019-sonicwall-cyber-threat-report/.

ST (2019). TF-A Overview [Online]. Available: https://wiki.st.com/stm32mpu/wiki/TF-A_overview.

TrustedFirmware Association (2019). Trusted Firmware-A Documentation [Online]. Available: https://trustedfirmware-a.readthedocs.io/en/latest/.

van der Veen, V., Fratantonio, Y., Lindorfer, M., Gruss, D., Maurice, C., Vigna, G., Bos, H., Razavi, K., and Giuffrida, C. (2016). Drammer: Deterministic rowhammer attacks on mobile platforms. *Proceedings of the 2016 ACM SIGSAC Conference on Computer and Communications Security*, pp. 1675–1689.

4

Accelerating Virtualized Distributed NVMe Storage in Hardware

Julian CHESTERFIELD[1] and Michail FLOURIS[2]

[1] *Sunlight.io Limited, Cambridge, United Kingdom*
[2] *Sunlight.io Greece, Heraklion, Greece*

Modern NVMe Flash devices deliver over 1 million random I/O operations per second (IOPS) at sub-100 µs latencies, making them orders of magnitude faster than traditional hard drives with rotating disks, especially for random I/O. Although the overhead of storage virtualization and distributed I/O management has also decreased, it has not kept up with the rapid decrease of NVMe flash latencies.

Currently, the time spent in storage virtualization and management per I/O operation has grown larger than the NVMe Flash hardware latency, and as a result, this introduces significant performance overheads. Moreover, this overhead usually affects host server CPUs running the distributed storage virtualization stack, thus limiting the CPU resources used by application workloads, essentially slowing them down.

To minimize these overheads and more efficiently unlock the raw hardware performance for applications, Sunlight.io has been building a hardware-assisted I/O path that leverages runtime-programmable FPGAs, to reduce the management overheads and drive NVMe storage access over the network at rates of up to 100 Gbps. At the same time, using an inline datapath in the FPGA enables the virtualized workloads to access network-attached storage with minimal latency.

For a color version of all figures in this book, see www.iste.co.uk/andrade/multi2.zip.

Multi-Processor System-on-Chip 2 – Applications,
coordinated by Liliana ANDRADE and Frédéric ROUSSEAU.

4.1. Introduction

For several years we have been witnessing the increasing migration trend of applications and workloads, traditionally run in on-premise servers to virtualized cloud infrastructures, where the application executes in virtual machines (VMs) or containers. However, there are still barriers for many organizations to migrate their applications to the cloud, the most important of which are performance and cost. It is well documented that the performance of an application executing in a virtual machine is usually significantly reduced compared to the same application running on an operating system (OS) without virtualization. This lost performance due to the virtualization overhead, is frequently referred to as the "virtualization tax".

The virtualization overhead not only reduces performance but can also significantly affect the cost aspect, especially for demanding applications. Maintaining high performance that is required for demanding applications incurs significantly higher costs for organizations for a cloud infrastructure with higher-end hardware, compared to baremetal servers with lower overheads.

Although performance and cost are still barriers in the migration of demanding applications to the cloud, limiting the growth of enterprise-level cloud computing, there are several other advantages in terms of cost, scalability and availability, that make migration to the cloud desirable.

In this chapter, we explain the challenges of high-performance cloud computing, especially with respect to storage I/O operations, and present our solution to accelerating virtualized distributed NVMe storage with hardware. Our work on the hypervisor layer, assisted by the virtualized hardware-assisted I/O datapath, allow an application running on a virtual machine to perform at the same level as a fast NVMe device directly attached to a baremetal server, without losing the scalability and flexibility advantages of a cloud platform. This is achieved by minimizing overheads and optimizing I/O operations at two main layers of a cloud infrastructure:

1. the hypervisor layer, where Sunlight.io has built a new hypervisor designed to minimize overheads for virtual I/O operation at the hypervisor level. This hypervisor which is called "NexVisor" is a key component for virtual I/O performance over both directly attached high-performance NVMe flash devices and disaggregated storage;

2. a disaggregated storage appliance that uses hardware mechanisms to allow virtual networked I/O operations to be executed by NVMe flash devices at low latency and great throughput. This appliance allows scaling of the NVMe storage capacity over a cluster of machines to large sizes that are not possible with host-attached devices on hyperconverged infrastructures.

In this chapter, we first provide a short introduction to virtualization and to the NVMe flash device technology. Then, we briefly discuss the concept of hyperconverged infrastructure (HCI) versus disaggregated cloud infrastructure, providing insights into the advantages of each type of setup. Then, we proceed with an overview of the NexVisor platform and explain how it has been designed to minimize I/O overheads and provide I/O performance that is equivalent to servers running baremetal OS.

4.1.1. *Virtualization and traditional hypervisors*

Virtualization of server hardware is a commonly used practice to provide scalable resource management, and it is essentially the enabling technology for cloud computing. There are many benefits of virtualizing hardware resources, primarily to enable efficient resource sharing (CPU, memory, network interfaces (NICs)) across a multi-tenant platform hosting a variety of operating systems (OSes). A hypervisor (HV) or virtual machine monitor (VMM) is a piece of software that creates, manages and runs virtual machines (VMs), or guest domains. The hypervisor presents the guest operating systems with an abstraction of a virtual operating platform and manages the execution of the guest operating systems. In addition to commonly deployed commercial hypervisors such as VMware, there are two dominant open-source hypervisor platforms: kernel-based virtual machine (KVM) (KVM 2020) and the Xen hypervisor (Linux Foundation 2020).

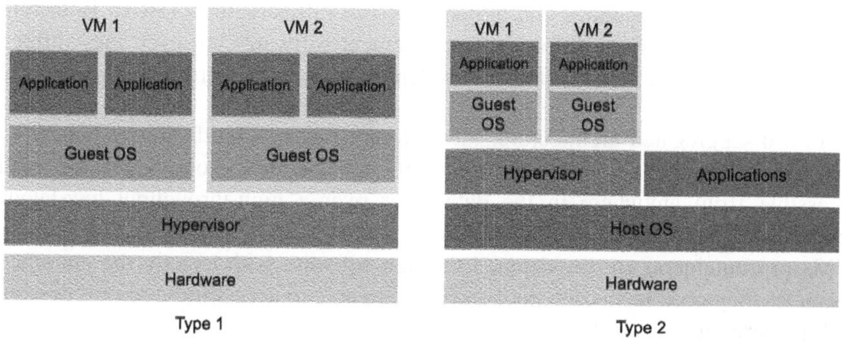

Figure 4.1. *Hypervisor types*

The Xen hypervisor provides a true type 1 hypervisor (Figure 4.1), in contrast to the KVM hypervisor platform. That is to say, that the hypervisor layer runs directly on the baremetal hardware, managing guest OS instances directly above it. There is no host operating system required and in this way, the type 1 architecture is considered to be a minimal, high-performance shim. In parallel to the virtualized guest systems running on the type 1 hypervisor, traditional Xen systems use a control domain, known as Dom0, which has privileged access to the hypervisor and is responsible for various tasks, including administering guest virtual machines, managing resource allocation for guests, providing drivers for directly attached hardware and offering network communication support. Guest virtual machines (DomUs) do not typically have direct access to real hardware, and thus all guest domain network and storage communication is managed through paravirtual device drivers, hosted in the control domain (Dom0), which in turn handles safe resource access by multiplexing the physical hardware.

For every guest VM paravirtualized (PV) device that is active, there is a corresponding driver in the control domain that allocates resources and handles the communication ring, over which, virtual IO requests are transmitted. The control domain is then responsible for mapping those IO requests onto the physical hardware devices behind them. Based on the Xen hypervisor platform, a guest OS can access a paravirtual IO interface using the standard paravirtualized (PV) driver, such as, Netfront for the Ethernet or Blkfront for the block storage device without any actual underlying hardware knowledge. In contrast to PV, HVM guests are fully virtualized, providing hardware support through their own unmodified drivers in the VM, while the hypervisor provides full support for hardware devices, either native or through hardware emulation.

4.1.2. *Hyperconverged versus disaggregated cloud architectures*

Virtualized cloud infrastructures are deployed in two types of architectures: disaggregated infrastructures and hyperconverged infrastructures (HCI). Both types share attributes in virtualization and management software; however, they differ in how wide their scope is, how hardware is deployed and what hardware is used.

In a disaggregated architecture, the different types of hardware resources (e.g. compute or storage) are separated into different resource pools, which are easily accessible from anywhere in the datacenter over a high-throughput low-latency network. In this manner, application workloads running in virtualized abstractions (VMs or containers) are provisioned through software with exactly the amount of resources they need from each pool and they are able to access these resources over a high-performance network.

Hyperconvergence is a different type of architecture where compute, memory, storage and networking hardware resources are combined into a single server system. Adding more of the same servers allows the scaling of all types of resources in parallel, i.e. adding the CPU, memory and storage resources of a server in the pool. Hyperconverged infrastructures result in reduced datacenter complexity and dependence on high-performance networking by exploiting locality (e.g. using storage devices local to the same server a VM is running on). HCI platforms rely on a hypervisor for virtualized computing, software-defined storage and virtualized networking, and they typically run on standard, off-the-shelf servers, resulting in lower costs. Multiple server nodes can be clustered together to create pools of shared compute and storage resources, designed for convenient consumption.

The differences in the two types of architectures are shown in Figure 4.2 and are summarized in the following subsections.

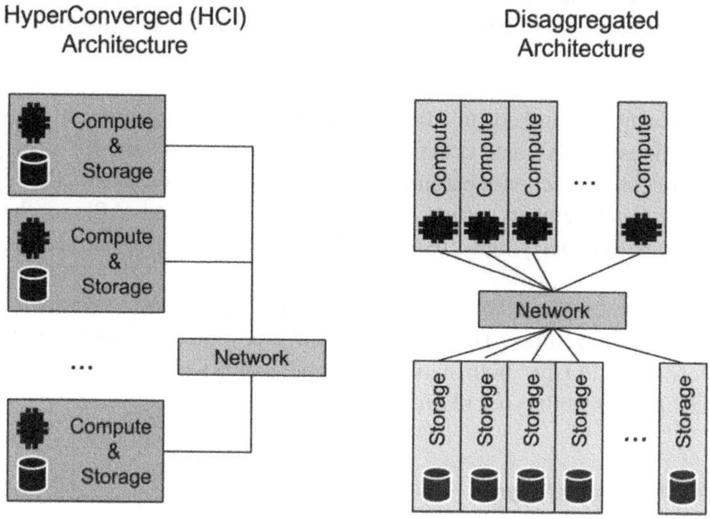

Figure 4.2. *Hyperconverged versus disaggregated architectures*

4.1.2.1. *Degree of resource virtualization*

HCI combines compute and storage resources onto one commercial off-the-shelf (COTS) box. These resources are virtualized and pooled across nodes controlled by the virtual storage appliance technology the HCI solution uses. These groups of nodes are called clusters.

Disaggregated infrastructure pulls apart as many resources as possible into separate devices dedicated to their own resource. Servers contain compute and memory hardware, whole sleds on a rack are for storage disks, and networking devices tend to be on the back side of the rack. The resources are virtualized and decoupled from the hardware they reside in, and thanks to the speed of the network fabric, the resource pool can be accessed from anywhere in the datacenter.

4.1.2.2. *Scalability of disaggregation and HCI*

When using a disaggregated infrastructure, there is more control over what resources are added when scaling is required. In HCI, storage and compute are restricted to the same box, so when one type of resource is added, the other resources (e.g. storage) come with it. This is not always reasonable, as sometimes the user only needs to add one type of resource (e.g. storage).

4.1.2.3. *Management software*

Both disaggregated infrastructure and HCI have management software designed to simplify and automate the datacenter's infrastructure. In both scenarios, the goal is to enable IT generalists, rather than distinct groups of specialists, to administrate the datacenter.

HCI software is linked to the hypervisor technology, which limits it to virtual machines (VMs) and containers. Management software for disaggregated infrastructure can interact with baremetal servers, VMs and containers. Disaggregated infrastructure management software also features a unified API so that it can communicate with underlying software and hardware from any other vendor.

For the purposes of our work on optimizing I/O performance, we see benefits in both types of architectures. The system we have designed and optimized provides high-performance I/O for both networking and storage traffic to and from tenant virtual machines, in both types of architectures.

4.1.3. *NVMe flash storage*

NVM Express (NVMe) (NVM Express, Inc. 2020) is a parallel high-performance interface and command set designed for non-volatile memory-based storage. NVMe devices have been designed to support a logical device interface where storage media is attached via the PCI Express (PCIe) bus, without the complexities and overheads of storage protocols such as SCSI or SATA (Broadcom 2020). Attaching high-performance flash solid-state drives (SSDs) over the NVMe interface allows I/O operations to be delivered with a very high data throughput (several GBytes per second). Moreover, modern flash devices can deliver over 1 million random I/O operations per second (IOPS) at sub-100 μs latencies over NVMe, making them orders of magnitude faster than traditional hard drives with rotating disks, especially for random I/O. For this reason, NVMe flash drives are the superior storage choice for data-intensive datacenters to obtain high-performance computing in cloud services (Peng *et al.* 2018; Barroso and Hülzle 2009).

As an example of a current NVMe storage, we consider a modestly priced consumer-based drive, using a 2- or 3-bit MLC NAND Flash technology on a PCIe Gen 3.0 x4, NVMe 1.3 interface. According to its manufacturer, it supports sequential read performance of up to 3,500 Mbytes/s and sequential write up to 2,700 MB/s. Its random I/O performance can reach up to 500,000 IOPS for read and write operations (Samsung 2020). This random I/O performance is orders of magnitude better than a hard drive and there are enterprise-level NVMe Flash drives that can achieve better performance than this.

4.2. Motivation: NVMe storage for the cloud

Although the overhead of storage virtualization and distributed I/O management has decreased in modern cloud architectures, it has not kept up with the rapid decrease of NVMe flash latencies. Unfortunately, even though the storage hardware performance has increased orders of magnitude with NVMe Flash drives, the real I/O performance for cloud applications has not been aligned with that increase in hardware performance. In both Hyperconverged (HCI) and disaggregated architectures, the software virtualization overheads have become the bottleneck in allowing cloud applications to follow hardware I/O performance evolution.

Motivated by this challenge of providing high-performance storage I/O operations to cloud applications, we have focused our work on minimizing virtualization overheads and optimizing I/O operations at two main layers of a cloud infrastructure:

1. the hypervisor layer, where we build a new hypervisor designed to minimize overheads for virtual I/O operation at the hypervisor level;

2. the disaggregated storage layer, where we design an appliance that uses hardware mechanisms to allow virtual networked I/O operations to be executed by NVMe flash devices at low latency and great throughput.

4.2.1. *Motivation for a new hypervisor*

Considering Xen as a reference type 1 hypervisor, we observe that its architecture for I/O (storage and network requests) is centralized. As shown on the right-hand side in Figure 4.3, in Xen, sending all the IO and network requests from guest domains through a central, core domain (Dom0) itself becomes a bottleneck that affects performance. The network and I/O performance in this architecture is limited, due to context switches, to the control domain (Dom0) that handles the network packet switching for all VMs. A new model is required to enhance and advance this architecture by making the packet forwarding and I/O layers as fast as possible, with slow path network functions offloaded to separate processing engines.

To address these limitations of a traditional type 1 hypervisor architecture (e.g. Xen), we have developed a new optimized hypervisor (shown on the left-hand side of Figure 4.3), which removes the centralized Dom0 model of a type 1 hypervisor and instead passes the logic and control to a higher-level software tool that can then communicate with the guest VMs. The core idea behind this approach is to remove any dependency on a local control domain (Dom0) for virtual machine setup, booting and resource allocation and instead, move this generic functionality into the hypervisor layer itself. More details on the design of the NexVisor are presented next in section 4.3.

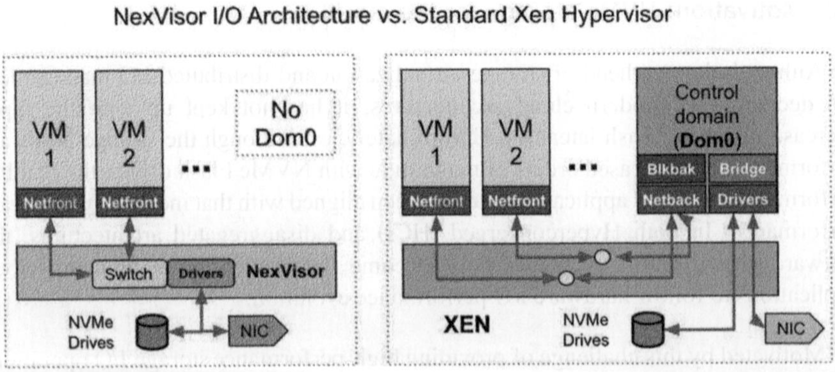

Figure 4.3. *Comparison of the NexVisor I/O architecture to standard Xen*

4.2.2. *Motivation for accelerating disaggregated storage*

File-based servers (NAS) or SAN servers, in general, use a general-purpose CPU, running an operating system such as a Linux distribution, with a network adapter and attached storage (NVMe/PCIe or disks). These servers virtualize the physical storage and provide software-defined storage abstractions such as virtual disks or files to the virtual machines (VMs), allowing flexible configuration and sharing of the resources. All processing and sharing of these virtual abstractions is performed by the software that runs on the generic CPU, which results in increased overheads when the I/O request rates increase. At the high IOPS rates of modern NVMe flash drives (i.e. millions of I/O requests per second), the storage virtualization overheads introduced by the software stack for I/O requests are significant and result in the "virtualization tax" that we have mentioned previously.

To address these overheads for disaggregated storage, we have designed and implemented an accelerated I/O appliance, which provides a high-throughput, low-latency network and disaggregated storage for the VMs on cloud infrastructures. In our I/O acceleration system, we keep providing virtualized software-defined storage (i.e. virtual disks) to VMs; however, at the data-path level, all networked storage requests and mapping of virtualized blocks to NVMe Flash are completely offloaded to an FPGA. Using a protocol based on the ATA-over-Ethernet (AoE) network storage protocol (Wikipedia 2020), all storage I/O requests in the form of Ethernet network packets are processed completely by the hardware logic, and the NVMe Flash drives are also attached and managed directly by our hardware logic. Thus, the network-to-storage connectivity and the virtual-to-physical translation are completely offloaded in hardware logic.

4.3. Design

To achieve our objective of accelerating virtualized distributed NVMe storage for high-performance cloud computing, we have focused on minimizing overheads and optimizing I/O operations at two main layers of a cloud infrastructure:

1. the hypervisor layer, where we have built a new hypervisor designed to minimize overheads for virtual I/O operations. The hypervisor layer is a key component for virtual I/O performance over both directly attached and disaggregated storage with high-performance NVMe flash devices;

2. the disaggregated storage layer, where we use hardware offloading mechanisms to allow virtualized networked I/O operations to be executed by NVMe flash devices at low latency and great throughput. Our design does not inhibit the virtualization flexibility of software-defined storage, but only accelerates the necessary I/O operations through the fast logic of the data processing path, which is executed on FPGA hardware.

In the rest of this section, we present our design in these two layers, the hypervisor and disaggregated storage layers.

4.3.1. *Optimizing the hypervisor I/O operations*

As discussed in section 4.2, the traditional hypervisor architectures (e.g. Xen or KVM) for I/O requests are centralized, sending all the IO requests from VMs through a central, core domain (Dom0 in Xen or host OS in KVM), which itself becomes a bottleneck that affects performance.

To address these limitations of a traditional hypervisor architecture, we develop a new optimized type 1 hypervisor platform named the "NexVisor". The NexVisor architecture, shown on the left-hand side of Figure 4.3, removes the centralized Dom0 model of a type 1 hypervisor and instead passes the logic and control to a higher-level software tool that can then communicate with the guests. The core idea behind this approach, is to remove any dependency on a local control domain (Dom0) for virtual machine setup, booting and resource allocation, and instead move this generic functionality into the hypervisor layer itself (see Figure 4.4).

4.3.1.1. *Design of the NexVisor hypervisor*

The main difference of the NexVisor from Xen and KVM is that there is no control domain (Dom0 in Xen terminology, or host OS for KVM). The centralized Dom0 model means that split driver requests have to pass through the control domain for each I/O request, which becomes a bottleneck, when many requests are being served to multiple guest domains (DomUs) at high rates, as can be supported by fast NVMe devices.

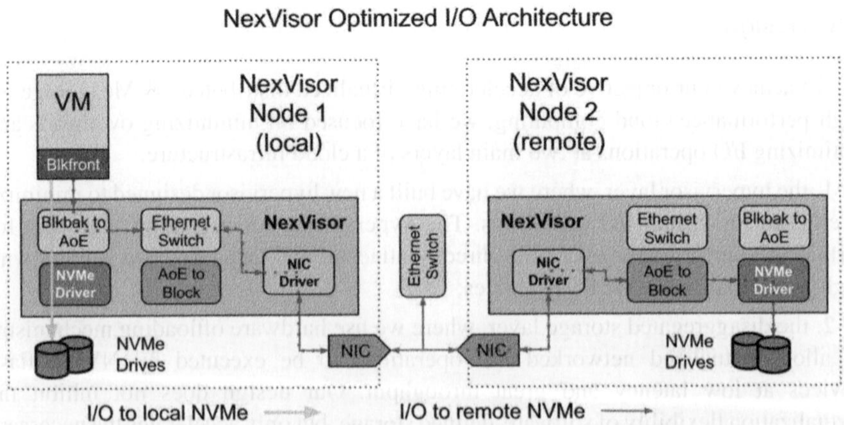

Figure 4.4. *Optimized I/O datapath operations in the NexVisor for local and remote NVMes over the network*

Next, we present the key elements in the design approach of the NexVisor:

1. **Hardware interface and drivers**: one of the issues of moving to a controller-less approach is that the drivers interfacing with the hardware, which were part of the control domain (Dom0), now need to be moved to another part of the hypervisor. In the NexVisor architecture, this is implemented by either:

a. creating lightweight driver domains (e.g. via a unikernel or minimal OS) that have access to the device drivers and the underlying hardware. In this case each piece of hardware has at least one driver domain associated with it, which the guests communicate with to access the physical resources;

b. integrating native device drivers for the hardware in the hypervisor layer itself. This has an even lower overhead for I/O devices; however, it is more complex to implement. In order for the NexVisor to provide the best performance for NVMe storage, such drivers for network interfaces and NVMe drives have been integrated in the hypervisor itself.

2. **Configuration and management**: due to the absence of a control domain (Dom0), the NexVisor architecture handles configuration and management significantly different from traditional hypervisors, such as Xen or KVM, which can use Dom0 (or host OS in KVM) for configuration and management services. To allow efficient management, the NexVisor architecture has moved a minimal essential set of tools from Dom0 to the hypervisor layer itself, adding support for control and configuration APIs needed.

3. **Management API**: in addition to moving configuration and control from Dom0 to the hypervisor, the NexVisor architecture features a remotely accessible (i.e. networked) control and management API, to accommodate the fact that there

is no control domain to run locally on each NexVisor. The result is an Ethernet-level command API that allows secure location-agnostic monitoring and the control of any hypervisor from a controller anywhere on the local Ethernet network. The NexVisor Ethernet-level API can be used in a distributed mode for monitoring and managing resources on multiple remote hypervisors (compute, network and storage resources) from a single controller, running on a VM anywhere in the local cluster.

4. **Monitoring API**: due to the absence of a control domain (Dom0), the monitoring information of the hypervisor and VMs is provided by the hypervisor layer. Each NexVisor sends this monitoring information periodically, or on demand, through Ethernet packets, which can be received, processed or stored by one or more monitoring agents anywhere on the local Ethernet network.

5. **Always networked and virtualized storage**: another important characteristic of the NexVisor architecture is that storage I/O requests cannot be sent in the usual block-back queues through Dom0 to storage drives, since there is no Dom0. Instead the NexVisor architecture, being designed specifically for distributed storage operation, sends the I/O requests either directly to the local (directly attached) device, or converts them into Ethernet frames and sends them via the ATA-over-Ethernet (AoE) network protocol, to the remote storage devices. Using Ethernet frames ensures that the overhead of TCP/IP processing is removed and as such the I/O virtualization overhead is significantly reduced. This design choice to always use storage as virtualized and networked, works well in combination with disaggregated storage, allowing the software-defined storage layer to flexibly configure and manage all storage volumes in the same manner, on local, directly attached storage drives, or on another host in an HCI cluster, or even on disaggregated storage providing networked I/O.

6. **Software-defined networking**: as an HCI platform, the NexVisor supports the creation and management of Ethernet-level virtual network overlays between nodes and VMs. An essential component for this support is an integrated internal switch, which is embedded in the hypervisor kernel, to handle Ethernet packet forwarding. This switch manages shared network and I/O resources from different NexVisor nodes, to provide aggregated network and storage resources that are linked and accessed from the guest VMs on each hypervisor.

Through the key elements described above, the NexVisor architecture allows much lower overheads and better I/O performance for both HCI and disaggregated infrastructures. Overall, the use of Ethernet control frames for the management, control, I/O and monitoring of the NexVisor platform, along with the driver domain and removal of Dom0, ensures that the virtualization and I/O layer overheads are much lower than the standard hypervisor implementations. This results in higher performance and better control for each of the resources and the guest VM domains, as shown in section 10.5.

4.3.2. *Design of accelerated disaggregated storage*

The Sunlight.io hardware I/O accelerated storage solution has a novel design (Figure 4.5), where the end-to-end storage system comprises a software-defined storage volume management architecture, while at the same time, allowing the acceleration of all the data processing requests to be handled entirely in the fast path of the storage FPGA hardware. The software-defined storage solution enables distributed logic to be applied across all elements of the communication system, providing the flexibility of the software-defined storage, while accelerating only the necessary operations through the fast logic of the data processing path.

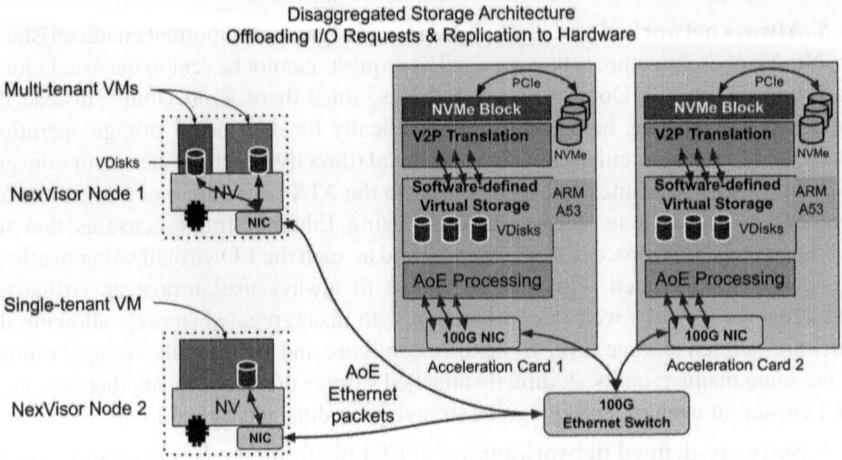

Figure 4.5. *High-level view of disaggregated storage architecture, showing the I/O requests originating from the VMs through the NexVisors and through the ethernet network to the acceleration cards*

The key elements of the software-defined accelerated storage design can be summarized in the following sections.

4.3.2.1. *Virtualization of physical storage drives*

A key feature is the physical disk partitioning and sharing of many virtual disks used by virtual machines (VMs). This is achieved through the concept of virtual to physical (V2P) block mapping tables. Each virtual disk replica in the disaggregated storage pool is described by a unique AoE virtual disk number (LUN), which has a corresponding V2P table, mapping the virtual to the physical blocks that actually reside on the physical storage. That way, we can partition and share the same physical drive between multiple client VMs. Furthermore, V2P tables allow us to span virtual disks among different physical drives.

4.3.2.2. *On-demand virtual disk management*

Another key feature supported by the software-defined storage layer of the NexVisor platform is the flexible management of virtual storage through the following operations:

1. create virtual disks (VDisks) of a specific size and redundancy (one or more replicas) over any physical NVMe Flash drive on the disaggregated storage pool. If more than one replicas are requested, the allocation manager selects available storage space on more than one storage drives (anywhere in the disaggregated storage pool) and reserves the capacity for the VDisk, saving the configuration state on persistent storage;

2. connect the virtual disk frontend onto a specific hypervisor to be used by a VM on that hypervisor. Once a VDisk is created and connected (i.e. the LUN is created and V2P table is mapped), the I/O request processing datapath is mapped from the frontend hypervisor through the AoE network I/O protocol, to the disaggregated storage backend nodes and there it is completely offloaded to hardware without any software processing running on a general-purpose CPU;

3. disconnect and/or reconnect the virtual disk frontend to a different hypervisor, following the VM in case of a migration. Using the AoE network protocol for the datapath mapping, allows easy migration and remapping;

4. delete virtual disks, which deallocates the physical space used by the data and removes all LUN and V2P mapping from the storage software-defined layers and the hardware acceleration state. This allows the storage space to be used by new VDisks and new VMs;

5. resizing existing virtual disks to a larger or smaller capacity. Growing the VDisk adds more capacity in its backend replicas, allowing the VM using it to store more data. Shrinking the VDisk is more difficult, since the VM must first shrink its filesystem or data volume and then instruct the storage layer to shrink the VDisk.

4.3.2.3. *Offloaded virtual disk replication*

Mirroring data to multiple replicas is essential for reliability and availability purposes; however, the data copy operations can introduce significant overheads. To minimize these overheads, we have extended the standard ATA-over-Ethernet (AoE) protocol to include a custom mirror command that is fully handled by the hardware. Specifically, a hypervisor holding a VDisk frontend can issue mirror commands to the backend nodes through the network to update VDisk replicas that reside on different physical drives and different hardware adapters or nodes, without the need to create duplicate AoE write requests to send to each replica. In this manner, the hardware handles the mirror command by issuing AoE writes to the remote replicas automatically, offloading replication from the frontend hypervisor and reducing its network load.

4.3.2.4. *Offloaded virtual disk copy*

The storage layer maintains redundant data replicas to allow availability in case of failures. After a failure, one or more replicas will become inconsistent and need to be re-synchronized with the latest live version of data from a live replica. This "replica repair" operation is performed by copying data from one replica to another and can incur overheads for data copy. To minimize these overheads and offload this operation to hardware, the client sends a "block copy" command to the hardware adapter, that holds the source replica with the data to be copied. Receiving that request causes the receiver to READ some blocks of data from the physically attached drive (depending on the V2P mapping) and then convert the READ response into a WRITE command, that is sent to a new replica endpoint as specified in the original "copy" command header. In this manner, a client can control and offload data copy operations to hardware, without having to read and process data itself.

4.3.2.5. *Offloaded virtual disk snapshot*

Virtual disk snapshots are essential management operations for creating backups and managing storage volumes. A hardware-based snapshot mechanism allows the hardware endpoint to create an efficient data snapshot of a virtual disk, without any significant overhead for the hypervisor. The client requests a new VDisk of the same size to be created (Snapshot child VDisk) and can then notify the hardware logic, via an atomic operation, to begin a snapshot relationship between the parent VDisk and the child VDisk. From this point in time, when new block WRITE requests are received for the parent VDisk, the hardware datapath checks a dirty bitmap for the child VDisk, to assess whether that logical block has been written on the child or not. If it has not, then the block WRITE is issued to the address allocated to the child VDisk, rather than the parent block address. Once the WRITE completes, the hardware logic then flips the address offset for the logical block between the parent and the child. It then updates the dirty bitmap for the child to record the fact that the child now stores a block of data for that logical address. In this manner, the child VDisk always stores a copy of the original data from the parent VDisk at the time of the snapshot creation. Furthermore, the snapshot child VDisk can be discarded atomically at any point in time without any impact on the parent VDisk, since there is no data dependency for the parent.

4.3.2.6. *Networked protocol*

4.3.2.6.1. ATA over Ethernet (AoE)

ATA over Ethernet (AoE) is a network protocol, designed for simple, high-performance access of block storage devices over Ethernet networks (Hopkins and Coile 2009). The standard AoE header format and bit-wise alignment is shown in Figure 4.6.

```
    0                   1                   2                   3
    0 1 2 3 4 5 6 7 8 9 0 1 2 3 4 5 6 7 8 9 0 1 2 3 4 5 6 7 8 9 0 1
   +-+-+-+-+-+-+-+-+-+-+-+-+-+-+-+-+-+-+-+-+-+-+-+-+-+-+-+-+-+-+-+-+
 0 |                   Ethernet Destination Addr                   |
   +-+-+-+-+-+-+-+-+-+-+-+-+-+-+-+-+-+-+-+-+-+-+-+-+-+-+-+-+-+-+-+-+
 4 |   Ethernet Destination Addr   |       Ethernet Source Addr    |
   +-+-+-+-+-+-+-+-+-+-+-+-+-+-+-+-+-+-+-+-+-+-+-+-+-+-+-+-+-+-+-+-+
 8 |                     Ethernet Source Addr                      |
   +-+-+-+-+-+-+-+-+-+-+-+-+-+-+-+-+-+-+-+-+-+-+-+-+-+-+-+-+-+-+-+-+
12 |     Ethernet Type (0x88A2)    |  Ver  | Flags |     Error     |
   +-+-+-+-+-+-+-+-+-+-+-+-+-+-+-+-+-+-+-+-+-+-+-+-+-+-+-+-+-+-+-+-+
16 |             Major             |     Minor     |    Command    |
   +-+-+-+-+-+-+-+-+-+-+-+-+-+-+-+-+-+-+-+-+-+-+-+-+-+-+-+-+-+-+-+-+
20 |                              Tag                              |
   +-+-+-+-+-+-+-+-+-+-+-+-+-+-+-+-+-+-+-+-+-+-+-+-+-+-+-+-+-+-+-+-+
24 |                              Arg                              |
   +-+-+-+-+-+-+-+-+-+-+-+-+-+-+-+-+-+-+-+-+-+-+-+-+-+-+-+-+-+-+-+-+
```

Figure 4.6. *ATA over Ethernet (AoE) header format (Hopkins and Coile 2009)*

Since AoE runs on layer 2 Ethernet, it is more comparable to Fiber Channel over Ethernet than iSCSI. Using fewer protocol layers, AoE is fast, lightweight and makes the protocol easier to implement on hardware logic, offering linear scalability with high performance. This is the main reason we have used it for our high-performance accelerated I/O platform.

To support the core commands required for offloading storage processing to hardware, we have extended the standard AoE frame layout with additional fields. These are described in the next section, presenting the networked storage operations.

4.3.2.6.2. Networked storage operations

Using the AoE network protocol, all storage I/O requests in the form of network packets, are processed completely by the hardware logic and the NVMe disks are also attached and managed directly by our hardware logic.

Network-to-storage connectivity is completely offloaded in hardware logic, as described in the previous section, where the key mechanisms accelerated. Specifically, the hardware handles the following six core operations:

1. **READ** – processing the information contained in the AoE packet header, the hardware logic reads the corresponding logical data offsets from the physical NVMe flash drive and returns the data as an AoE response to the client. The logical-to-physical block translation is performed using the LUN and V2P tables described in section 4.3.2.7;

2. **WRITE** – processing the information contained in the AoE packet header, the hardware logic writes the attached payload data to the logical data offset, mapped to the physical NVMe Flash drive and returns a completion response. The logical-to-physical block translation is performed as in the READ command handling;

3. **WRITE_MIRROR** – extended header of the WRITE command. The logic duplicates the received WRITE packet and forwards it on to the destination node as signaled in the header. This operation enables the client (hypervisor) to avoid copying data in memory and transmitting multiple frames with duplicate data over the network, to different endpoints for replication. Essentially the replication operation is offloaded to the hardware logic, which allows the client to use those CPU cycles and memory for processing;

4. **COPY** – the hardware logic reads the data from the physical drive at the offset indicated in the packet header (mapped via V2P tables) and forwards the response as a WRITE command to the specified destination node as signaled in the header. The destination node sends a completion response to the client once it has completed the WRITE operation, so the operation is fully acknowledged. This operation enables the client to avoid sending a READ request, processing the response and then transmitting a WRITE request to the destination.

The COPY command is used for resynchronizing replicas. It is sent to the master (live) replica and acts as a READ command for the live data and is then forwarded to the target slave (failed) replica where it functions as a normal WRITE command (without the sequence number). The resynchronization process is managed by the client process, which initiates all COPY commands.

5. **CHECKPOINT** – the client sends a unique generation checkpoint via the AoE protocol to all destination nodes, to ensure data consistency across replicas of the same VDisk. Any node that is unresponsive will not generate the update checkpoint and will be determined as failed, according to the consistency algorithm applied by the replication logic;

6. **SNAPSHOT** – the client sends a command via an AoE packet to a destination node to trigger an atomic content snapshot. The hardware-based snapshot mechanism allows the creation of an efficient data snapshot of a virtual disk without any significant overhead for the hypervisor. The detailed description of the operation is presented in section 4.3.2.5.

These commands are generated by the hypervisors, that serve the I/O request queues of the virtual machines, as shown in Figure 4.5. On the receiving end of these commands are the hardware processing units (i.e. the FPGAs), that are receiving the AoE packets through the network and their logic controls access the directly attached NVMe flash storage drives. The FPGA units are capable of translating commands received as AoE network packets into IO operations that are directly issued to the drives at line rate. However, one of the key aspects of this work is that the software control logic can communicate a virtual LUN table to the FPGA, to enable it to provide a virtual storage management overlay function on top of the raw physical storage.

The key benefit of the client (hypervisor) being able to offload to the FPGA processing unit, these core operations, is that through this design the hardware logic becomes stateless. This means that every AoE packet received can be processed independently of any other, and the responsibility for data coherency, snapshot management and mirroring is placed on the software-defined frontend and endpoints, while all the heavy processing is entirely managed in the fast accelerated datapath at line rate.

4.3.2.7. *Storage virtualization*

One of the essential properties of our design is that the system allows software-defined storage through virtualized abstractions. The system allows the flexibility of creating and managing virtual disks, that are provided to virtual machines for storing their data. The software-defined storage features that are supported, depend on the following mechanisms provided by the accelerated datapath, in order to provide flexible virtualized storage abstractions, which offload their virtual-to-physical translation to hardware units. This is essential for the fast datapath acceleration. The mechanisms provided for this purpose, shown in Figure 4.7, are described next.

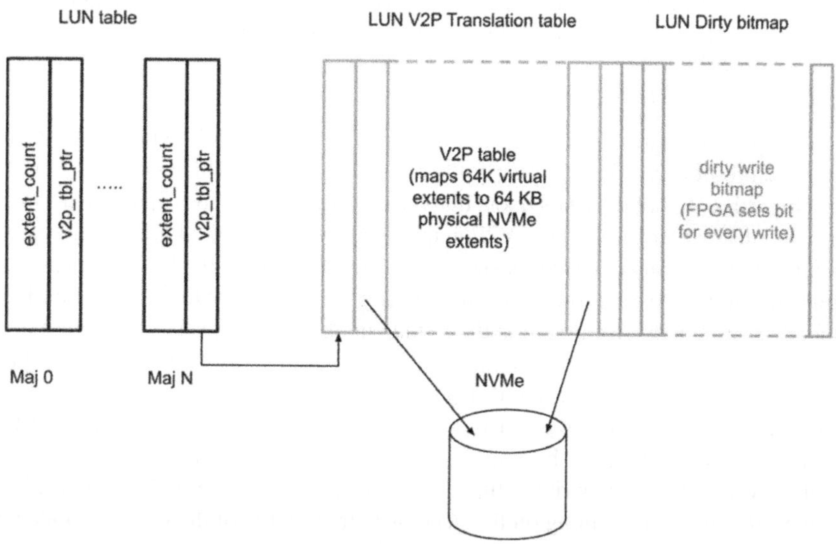

Figure 4.7. *Storage virtualization data structures used by the accelerated datapath*

4.3.2.7.1. LUN table

To manage the valid virtual disks (LUNs in the local optimized FPGA datapath), the hardware logic keeps a LUN table in its internal memory, which will contain one entry per LUN (VDisk). Assuming the maximum number of available LUNs is 64 K (since the AoE major number is 16 bits), the table can have a total of 64 K indexed entries. Each entry contains information about the size of the LUN, the "chunk size" used for the dirty bitmap and a pointer to the LUN's V2P table address.

4.3.2.7.2. Virtual-to-physical translation table structure

To manage virtual-to-physical storage block translation for each virtual disk (LUN), where virtual blocks are mapped as AoE LUN block addresses, and physical blocks are mapped on the NVMe block storage addresses, the accelerated (FPGA) datapath will keep a virtual-to-physical (V2P) table in memory. In order to fit in internal memory and avoid higher DRAM latencies, the size of the table is kept small using a predefined "chunk size" for contiguous physical blocks.

The structure of the V2P table uses physical addresses in the [0...*nvme_capacity*] range, indexed by a virtual chunk address. This chunk address maps 16 MB of the virtual block address space to 16 MB of physical storage (NVMe) space. Each logical chunk entry must have a 32-bit value that is non-zero, to indicate that the chunk is allocated on physical storage. If it is not allocated (i.e. has a value of zero), then the FPGA logic passes the AoE write request to the slow path software logic for processing. If it is a READ request for a block that is not allocated (has a value of zero), then the FPGA logic generates a response packet with zero data and responds back to the AoE client without needing to access the block device.

4.3.2.7.3. Dirty extent bitmap

As shown in Figure 4.7, the last required structure, directly following the V2P table data structure of each LUN, is a dirty logical extent bitmap, which contains 1 bit for every logical 64 KB of the LUN. Note that the dirty extent bitmap is updated in a much finer granularity compared to the V2P mapping. The number of 64 KB extents can be trivially calculated since the chunk size, extent size and size of the LUN (expressed in chunk size units) are all known and static during the lifetime of the LUN. The FPGA datapath is responsible for updating the bitmap in memory, on every completion of an AoE write request, in order to maintain an in-memory record of which extents have been written at least once (i.e. are dirty) since the bitmap was last zeroed. If a bit for a specific extent of the LUN is zero, then this means that there was no data written to that extent since the last "bitmap reset" time. If the bit is set to one, this implies that there was at least one write to that extent (on any part of the 64 KB extent). Note that the software tools running on the general-purpose CPU of the backend node will periodically read and reset the dirty bitmap, whenever a checkpoint event is triggered across the data mirrors. The dirty bitmap is used to detect which extents have modified data since the last checkpoint (i.e. when the bitmap was reset).

4.4. Implementation

We have implemented the design presented in the previous section for both the hypervisor layer and the disaggregated storage appliance. The implementation details are presented in this section.

4.4.1. *The NexVisor platform*

We have implemented the NexVisor hypervisor design for x86-64 architectures, as described in section 4.3.1. Hardware drivers for network interfaces and NVMe drives have been integrated in the hypervisor layer. We have also implemented a controller domain as a VM, which can be executed on any hypervisor in a local network, or migrated to any other hypervisor in the cluster in case of failure. The controller stack can control all the hypervisors in the cluster through the Ethernet-level management API. The controller provides command-line tools and a web-based user interface (UI) to allow users to configure software-defined networking and software-defined storage, as well as reliability and performance policies for the application of virtual machines.

4.4.2. *Accelerated disaggregated storage*

The Sunlight.io Acceleration Card provides a high-throughput, low-latency network and disaggregated storage appliance, which can be combined with the Sunlight.io NexVisor cloud infrastructure platform.

The implementation supports multiple $N \times NVMe$ physical drives and will present them to the software-defined storage layer as $N\times$ separate block storage devices (`/dev/disk1`, `/dev/disk2`, ...), where virtualized volumes can be created for VMs with specified redundancy policies (e.g. replication, locality). In the current implementation, each virtual disk has its own assigned LUN and corresponding V2P table, which maps the whole VDisk in 16 MB chunks. The LUN entries reside in the *LUN Table* BRAM inside the FPGA PL (programming logic), whereas the V2P tables, which are dynamically allocated, reside in the DDR memory of the PS.

4.4.2.1. *Hardware specification*

The Sunlight.io Acceleration card in our current implementation provides 2x 100 Gbps Ethernet ports and 2x NVMe M.2 slots along with hardware offloaded storage acceleration logic.

Specifically, the hardware supports the following features:

– 1x PCIe Gen3 16x-Lane for host interface;

– 2x 100 Gbps QSFP28 ports;

– 2x PCIe Gen3 4x-Lane M.2 slots for NVMe drives;

– Xilinx Zynq UltraSCALE+ FPGA in a SoC;

– ARM-64 A53 control CPU embedded in the FPGA SoC;

– 2 x 16 GB SoDIMM RAM.

4.4.2.2. *Architecture*

The Xilinx Zynq SoC's internal system interfaces for (FPGA-based) Ethernet traffic (including AoE) are AXI4 Streams with 512-bit data field, running at 250 MHz, allowing a peak BW of 16 GB/s between the host-SoC PCIe (Gen3 x16 lanes) DMA interface. The AoE subsystem (and other FPGA-based IPs) can also communicate with the SoC's A53 cores via DMA. Note that the A53's AXI HP slave interfaces have a maximum data width of 128 bits, allowing at most 4 GB/s bandwidth (at 250 MHz). Configuring the Zynq UltraSCALE+ IP block with the default AXI HP width of 32 bits will result in a peak BW to the DMA core of only 1 GB/s.

The NVMe host core has proprietary interfaces for submitting NVMe access requests and gathering access completion notifications, while payload data passed directly between the AoE subsystem cores and the NVMe drive's PCIe (Gen3 x4 lanes) bridge (to root complex) interface (AXI w/ 128-bit data 250 MHz) through dual-port URAM buffers, allows a peak throughput of 4 GB/s to/from the SSD. The data buffer ports facing the AoE subsystem are 512-bit data @ 250 MHz, allowing payload data to be transferred at 16 GB/s from/to the AoE IP blocks.

The AoE "fast path" processing is handled by several IP blocks, as shown in Figure 4.8. The interfaces and operations of these IP blocks are described next.

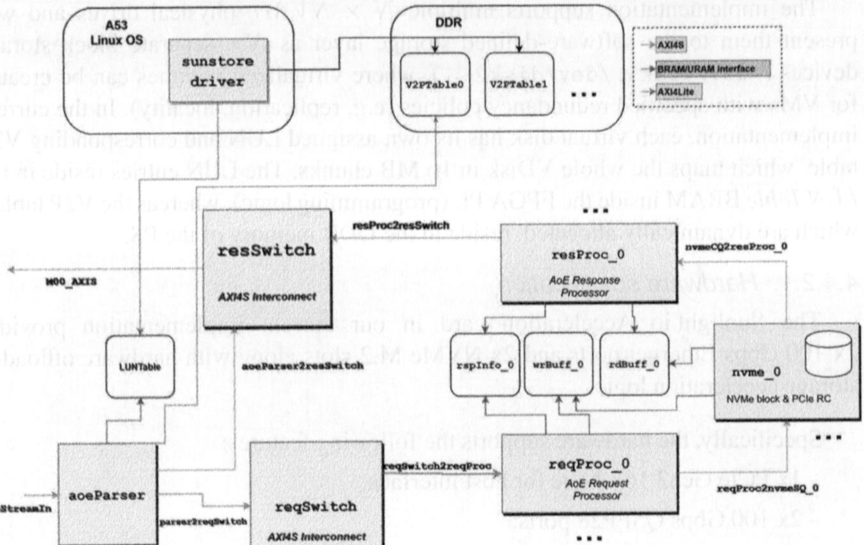

Figure 4.8. *Hardware architecture for the disaggregated storage acceleration card*

All incoming network traffic hits the aoeParser block, which is the frontend of the AoE acceleration subsystem. The aoeParser block extracts some basic info and forwards the AoE packet to the appropriate request processor. All request processors send their responses back to the response switch (resSwitch), which is responsible for sending the network traffic to the networking logic of the system. The interconnects between those components are based on AXI. In the block diagram in Figure 4.8, the green arrows indicate an AXI4 Stream interconnect, the red arrows indicate an BRAM/URAM interconnect and the blue arrows indicate an AXI4 Lite interconnect used primarily for communication with the processing CPU (A53) and DDR memory in a cache-coherent manner.

The role and operations in each hardware logic block are:

1. `aoeParser`: receives the incoming Ethernet AoE frame and extracts, source and destination MAC addresses, AoE command type, ATA command type. If the command is READ, WRITE, MIRROR or COPY, it extracts the major/minor numbers, the LBA and the number of sectors. It reads the fast BRAM LUN Table at position described in major number and gets the physical address of the corresponding V2P table in the DDR memory. It reads the V2P table in the DDR and gets the 16 MB physical chunk offset and disk ID based on the specified LBA. At the end, it forwards the packet to the AXI Stream switch;

2. `reqSwitch`: an AXI Stream interconnect switch to multiplex traffic between the different request processors and disks. The switching is done based on the disk ID extracted previously from the V2P table;

3. `reqProc_0`: receives an incoming AoE request and translates it into an NVMe command. In order to communicate with the NVMe interface logic, it makes use of three URAM buffers used as rings. The rspInfo block stores information about the request, such as MAC addresses in order to be able to generate the corresponding response. `wrBuff` is used for storing the payload data for WRITE commands. The NVMe access block reads the data from that buffer and accesses the physical disk. rdBuff is used for storing the disk data responses for READ commands;

4. `nvme_0`: this is a proprietary NVMe interface IP block, handling the PCIe Root Complex and the underlying NVMe physical drive. It converts the custom wire commands to NVMe-compliant commands;

5. `resProc_0`: this block is a response processor. Receiving the responses from the `nvme_0` block, it translates them to the proper network packets to be sent back to the client. It receives the network information from the `rspInfo` buffer and the responses for WRITE/READ commands from the `wrBuff/rdBuff` buffers;

6. `resSwitch_0`: an AXI4 stream interconnect switch which multiplexes traffic from the different response processors and sends them to the main FPGA switch logic, based on the destination MAC address;

7. `sunstore driver`: a Linux kernel module driver, running on the ARM A53 CPU. It configures all AoE-related blocks and the NVMe block, also providing the interface between the software-defined storage management stack and the hardware acceleration logic. It provides mechanisms for the mechanisms described in sections 4.3.2.1 and 4.3.2.2, such as creating/deleting VDisks, and managing their V2P tables dynamically at runtime.

Each NVMe disk is accompanied with its own blocks for `nvme_N`, PCIe host, request processor (`reqProc_N`), response processor (`resProc_N`) instances and three buffers (`wrBuff`, `rdBuff`, `rspInfo`).

4.4.2.3. *Storage node*

Storage management and virtualization services for the local NVMe drives are provided by software running on the Linux OS of the A53 ARM CPU of the card. This storage node also runs the *sunstore* driver, which is responsible for maintaining the LUNs and V2P tables. This driver will create the appropriate *LUN Table* entries and the V2P tables when it is probed by the storage management software. Besides this driver, distributed storage management tools on this storage node discover and communicate with other storage nodes in the local network, creating a distributed storage layer, where the NexVisor platform creates VDisks and stores the data of the VMs in a flexible manner.

4.5. Results

To measure the performance results of our optimizations, we performed several measurements, summarized in this section.

4.5.1. *Sequential reads*

In this experiment, we have used VMs running fio on four NexVisor nodes. Each of the 4 NexVisor hosts had one 10 Gbps link, so the total throughput of four clients was 40 Gbps. The fio benchmark, on each client VM, was performing sequential reads on a separate NVMe drive on the disaggregated acceleration card. The physical NVMe drives have been "trimmed" (i.e. flash blocks were discarded), in order to provide the maximum possible read IO per second (IOPS) performance. The TRIM operation on an NVMe drive essentially resets the garbage collection algorithm, and the drive can respond faster. The purpose of this measurement was to see if the acceleration card can saturate the high-performance NVMe drives through requests issued by fio on the NexVisor VMs, and over the 10 Gbps network links with AoE.

As shown in Figure 4.9, the acceleration card always achieved the theoretical maximum of the line rate, saturating the network links with different NVMes. We also observe that the card has achieved perfect scaling up to 1,200 K IOPS with four clients.

4.5.2. *Sequential writes*

This experiment is similar to the sequential read. We have used VMs running fio on four NexVisor nodes and doing sequential writes. Each of the four NexVisor hosts had one 10 Gbps link, so the total throughput of four clients was 40 Gbps. The fio benchmark, on each client VM, was performing sequential writes on a separate NVMe drive on the disaggregated acceleration card. The purpose of this measurement was to see if the acceleration card can saturate the high-performance NVMe drives through requests issued by fio on the NexVisor VMs and over the 10 Gbps network links with AoE.

As shown in Figure 4.10, the acceleration card always achieved the theoretical maximum of the line rate, saturating the network links with the different NVMes.

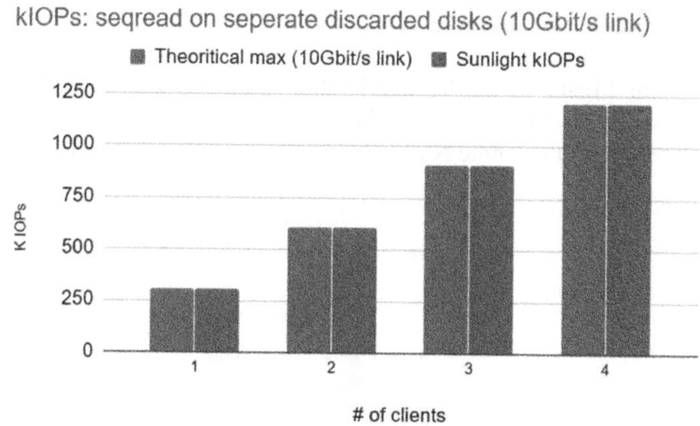

Figure 4.9. *Thousands of sequential read I/Os per second, using fio on four VM clients, each reading on a separate NVMe drive*

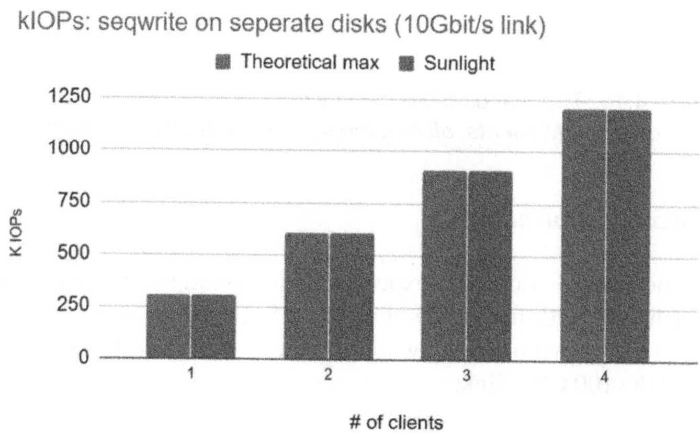

Figure 4.10. *Thousands of sequential write I/Os per second, using fio on four VM clients, each writing to a separate NVMe drive*

4.5.3. *Sequential reads on one NVMe drive*

In this experiment, we have used only one NVMe drive to demonstrate the saturation of that drive. We used VMs running fio on four NexVisor nodes, each of which had one 10 Gbps link, so the total throughput of four clients was 40 Gbps. In this case, the fio benchmarks on all four client VMs were performing sequential reads on the same NVMe drive on the disaggregated acceleration card. The physical NVMe drive has been "trimmed" to provide the maximum possible read IO per second (IOPS) performance. The purpose of this measurement was to see if the acceleration card can saturate the NVMe drive through requests issued by fio on the NexVisor VMs and over the 10 Gbps network links with AoE.

As shown in Figure 4.11, the acceleration card has achieved the maximum 700 K IOPS of the NVMe drive with three clients. The network links in the case are not saturated, since the single NVMe limits the overall performance.

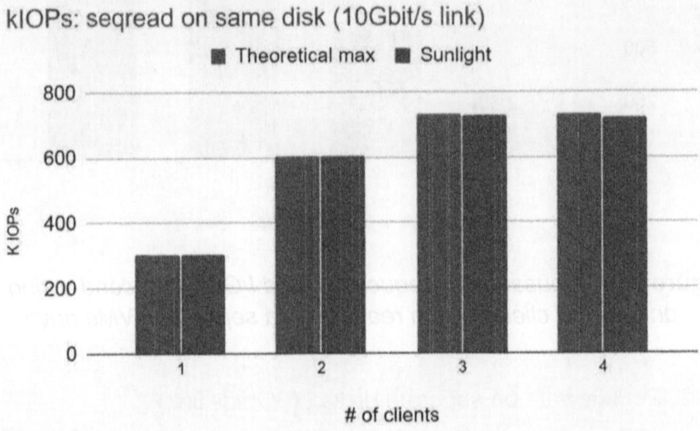

Figure 4.11. *Thousands of sequential read I/Os per second, using fio on four VM clients, all reading from the same NVMe drive*

4.5.4. *Network performance*

We have performed additional experiments to investigate the limits of network throughput achieved with the acceleration card. In the setup of this experiment, we have used two servers with two network interfaces (NICs) of 40 Gbps each, which should exceed the 100 Gbps link.

As shown in Figure 4.12, we have achieved $\tilde{8}6$ Gbps using an AoE request generation utility on separate NVMe drives. With the same utility for AoE request generation, we have achieved 104 Gbps when responding to AoE requests directly from the hardware logic, without doing actual NVMe I/O.

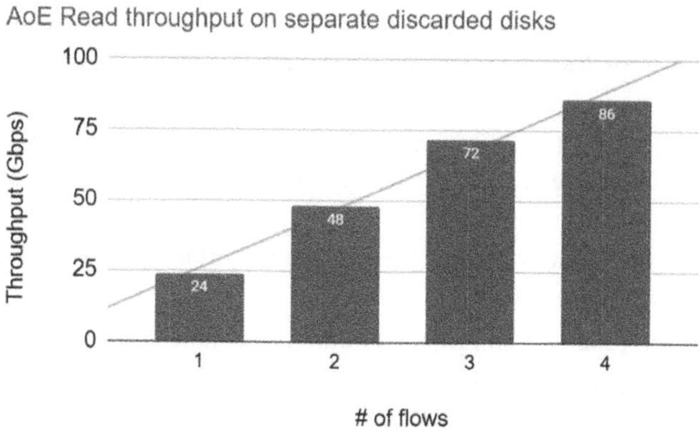

Figure 4.12. *AoE read throughput scaling for one to four client flows*

4.6. Conclusion

In this chapter, we have explained the challenges of high-performance cloud computing, especially with respect to storage I/O operations for demanding applications. We have presented our solution to accelerating virtualized distributed NVMe storage with hardware, as well as our work on the hypervisor layer.

Through this work, the virtualized, hardware-assisted I/O datapath of an application running on a virtual machine performs similarly to the I/O of a fast NVMe device, directly attached to a baremetal server, but without losing the scalability and flexibility advantages of a cloud platform. This is achieved by minimizing overheads and optimizing I/O operations at two main layers of a cloud infrastructure:

1. the hypervisor layer, where Sunlight.io has built the NexVisor, designed to minimize overheads for virtual I/O operation at the hypervisor level;

2. the Sunlight.io acceleration card, a disaggregated storage appliance that uses hardware mechanisms to allow virtual networked I/O operations to be executed by NVMe flash devices at low latency and high throughput. This appliance allows the scaling the NVMe storage capacity over a cluster of machines to large sizes that are not possible with host-attached devices on Hyperconverged Infrastructures.

Through this work, Sunlight.io has efficiently unlocked the raw hardware performance for virtualized applications and minimized infrastructure costs for businesses.

4.7. References

Barroso, L. and Hülzle, U. (2009). *The Datacenter as a Computer: An Introduction to the Design of Warehouse-Scale Machines*. Morgan & Claypool Publishers.

Broadcom (2020). Storage and PCI Express – A Natural Combination [Online]. Available at: https://www.broadcom.com/solutions/enterprise-storage/storage-and-pcie.

Hopkins, S. and Coile, B. (2009). AoE (ATA over Ethernet) [Online]. Available at: https://web.archive.org/web/20161025044402/, http://brantleycoilecompany.com/AoEr11.pdf.

KVM (2020). Main page – Kernel Virtual Machine [Online]. Available at: https://www.linux-kvm.org/.

Linux Foundation (2020). Xen Project [Online]. Available at: https://xenproject.org/.

NVM Express, Inc. (2020). Specifications – NVM Express [Online]. Available at: https://nvmexpress.org/specifications/.

Peng, B., Zhang, H., Yao, J., Dong, Y., Xu, Y., and Guan, H. (2018). MDev-NVMe: A NVMe storage virtualization solution with mediated pass-through, *Proceedings of the 2018 USENIX Annual Technical Conference*.

Samsung (2020). Samsung SSD 970 Pro [Online]. Available at: https://www.samsung.com/semiconductor/minisite/ssd/product/consumer/970pro/.

Wikipedia (2020). ATA over Ethernet [Online]. Available at: https://en.wikipedia.org/wiki/ATA_over_Ethernet.

5

Modular and Open Platform for Future Automotive Computing Environment

Raphaël DAVID[1], Etienne HAMELIN[1], Paul DUBRULLE[1], Shuai LI[1], Philippe DORE[1], Alexis OLIVEREAU[1], Maroun OJAIL[1], Alexandre CARBON[1] and Laurent LE GARFF[2]

[1] *LIST, CEA, Gif-sur-Yvette, France*

[2] *System Engineering Division, Renault, Guyancourt, France*

With the advent of connectivity services, cars are entering our digital world. Among the many consequences of this technological progress, the new mindset of customers towards their car is a true game changer. As the vehicle is becoming smarter, we expect it to always be up to date and offer the latest features and technologies. This versatility is however out of reach when relying on current electrical and electronic (EE) architectures, inherited from 35 years of incremental innovations. CEA, in cooperation with the Renault–Nissan–Mitsubishi Alliance, has developed a modular and real-time computing platform to face the challenges of future connected and automotive vehicles. Through the integrated design of an innovative hardware platform, embedded software and tools, we show that we are in a position to open this platform to software providers and manage the cohabitation of low-criticality applications with the critical operational software of the car. We leverage on an innovative model of computation and communication to express and formally analyze a behavioral contract for the interactions between functions. Embedded monitors can then verify at runtime that this contract is satisfied and switch to degraded mode in

For a color version of all figures in this book, see www.iste.co.uk/andrade/multi2.zip.

Multi-Processor System-on-Chip 2 – Applications,
coordinated by Liliana ANDRADE and Frédéric ROUSSEAU.

case of deviation. In this chapter, we come back retrospectively on the design process that allowed us to revolutionize the way car makers will design EE architectures for future smart vehicles. We cover the three pillars of this work to highlight key features of the solution from the point of view of hardware, software and tools. Finally, we conclude the chapter with an analysis regarding the mutation of the supply chain and the way new players may integrate the automotive industry.

5.1. Introduction

The mobility industry is shaken by the advent of artificial intelligence and connectivity that paves the way for new applications and uses of mobile systems. The lack of agility and efficiency of the electrical and electronic infrastructure is, however, a barrier to the adoption of these new technologies.

Automotive electrical and electronic (EE) architectures are inherited from more than 35 years of incremental innovations where each feature upgrade leads to the addition of a new computer (ECU) in the car. The resulting architectures are known to be very reliable, and they sustain a structured design process with a supply chain minimizing production risks. However, with up to 100 ECU in high-end cars, these architectures cause ever-increasing integration concerns and fail to leverage the whole value of data produced by car sensors (Traub *et al.* 2017).

In this old industry, Tesla has shown as early as 2013 that reconsidering EE architectures around a centralized computing solution allows us to disconnect electrical concerns from the development of car features. This was a groundbreaking step to revisit the user experience, leveraging on a flexible infrastructure, opening the way to SW personalization and upgrades all along the vehicle life cycle (Obergfell *et al.* 2019a).

Large suppliers are now also moving towards electronic aggregation in domain controller units (DCU) that gather multiple features inside a specific multi-purpose computer (e.g. for ADAS, cockpit or multimedia purpose). This trend is highlighted in McKinsey and Company (2019) and illustrated by Figure 5.1. The embedded electronic revolution for cars has begun, and this market will sustain a 7% CAGR to reach 43% of cars ECU at the 2030 timescale.

The Renault–Nissan–Mitsubishi Alliance and CEA anticipated this mutation in 2016 and started the FACE co-development project. FACE stands for Future architecture for Automotive Computer Environment. The project goal is to go one step further in centralization, with the radical approach to aggregate all car features in one single place, targeting the whole range of vehicle segments of the Alliance and opening the computing platform to the automotive community and attractive newcomers bringing breakthrough ideas in car usage modality.

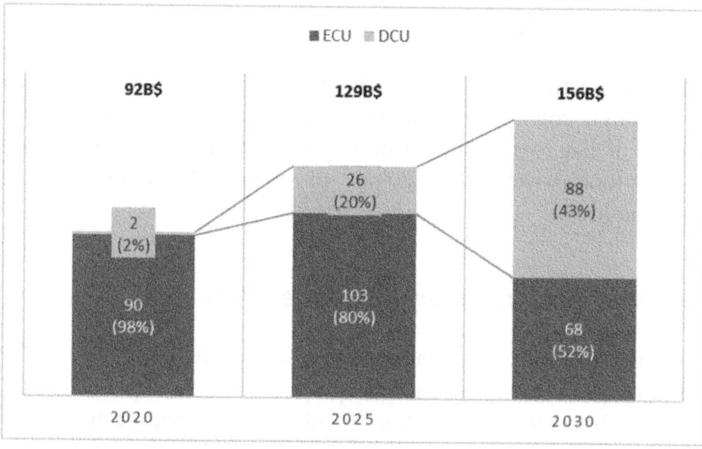

Figure 5.1. *Overall ECU/DCU costs (B$) evolution and breakdown between standard ECU and DCU from 2020 to 2030 (McKinsey and Company 2019)*

This early move is well positioned in the current trends to answer the tough challenges of future connected and autonomous vehicles (Vdovic *et al.* 2019). In this chapter, we detail how we addressed what we consider to be the most prominent challenges:

– How should we address automotive electronics diversity and performance range, from entry models to high-end connected and autonomous cars?

– How should we support the diversity of SW application needs that will allow us to capture the whole range of innovative applications by minimizing automotive electronics entry tickets?

– How should we safely manage the integration of multi-source features and address the resource sharing and resulting integration concerns all along the SW life cycles?

Those challenges have been tackled with a holistic approach taking jointly in consideration hardware, software and tooling concerns. Section 5.2 describes the foundations of the FACE solution. System architecture is first discussed from the hardware point of view, before we discuss the embedded software infrastructure guaranteeing efficient and safe coordination of vehicles features. Finally, we explain the approach taken on the tooling side to safely manage system-level design and application integration. Section 10.5 depicts the implementation results of all these concepts, again covering hardware, embedded software and tooling. We demonstrate the benefits of the FACE approach in section 5.4 through the evaluation of an adaptive braking system. Finally, we conclude the chapter by summarizing the FACE project impact, from both technical and supply chain evolution perspectives.

5.2. Outline of this approach

The FACE project revolves around three pillars: first, the centralization of all functions on a single common computing platform; second, a modular approach with the integration of heterogeneous hardware and software components; and finally, an end-to-end, tool-assisted methodology to support the application design and integration process.

5.2.1. *Centralized computation, distributed data*

Following the trend already mentioned of disconnecting electrical concerns from car features development, the FACE project proposes a new computation infrastructure based on a centralized approach. Contrary to other existing approaches (Reinhardt and Kucera 2013; Sommer *et al.* 2013; Traub *et al.* 2017), this infrastructure is not limited to domain centralization. The FACE architecture relies on a single modular solution targeting the whole vehicle range of the Alliance, embedding all the features on a single platform. The cohabitation between functions of mixed criticality is ensured by the software infrastructure presented in section 5.2.2.

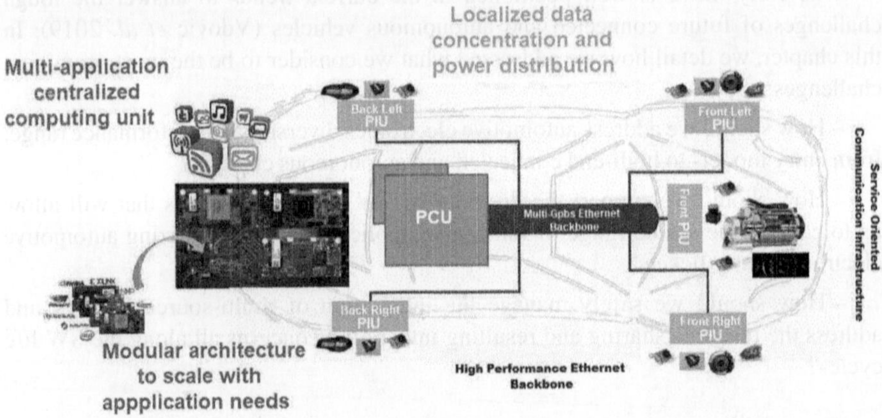

Figure 5.2. *Overview of the FACE PCU and PIU infrastructure*

The hardware infrastructure depicted in Figure 5.2 is composed of a centralized computing unit, the physical computing unit (PCU), hosting all car features. The functions hosted on the PCU can have access to all data produced by the vehicle's sensors thanks to physical interface units (PIU) distributed across the vehicle, acting as dedicated smart gateways. The PIUs can in turn receive commands from the functions hosted on the PCU to forward them to the vehicle's actuators. All these accesses are obviously subjected to access control for security reasons, but the idea is that if a function has the right and the need to access some sensor data, it is available to it, regardless of when the function is integrated on the PCU (early design phase or extension during product life cycle).

To support the transport of sensor data and commands, a high-bandwidth Ethernet network connects the PIUs and the PCU, implementing the time-sensitive networking (TSN) protocol suite to schedule packets to time-critical applications and ensure deterministic communication.

5.2.2. *Modularity and heterogeneity*

The functionalities offered by connected and autonomous vehicles rely on a large range of applications, presenting multiple computing patterns. Moreover, the automotive industry presents multiple segments from microcars to full-size luxury ones. These multiple segments present different functionalities and thus have different needs in terms of computing platforms. In order to serve these needs, and efficiently process the different underlying applications, the PCU is designed as a modular and heterogeneous computing platform, both at the hardware and software levels. The following sections present the main concepts of this approach.

5.2.2.1. *Modularity and heterogeneity at the hardware level*

At the hardware level, the PCU is a generic motherboard, readily extensible through the addition of daughterboards called computing sub-systems (CSS). The motherboard runs the computational payloads that are common to all automotive segments (e.g. body dynamics, driving assist, cockpit, etc.), while CSS provide the necessary modularity in computing power in order to achieve a scalable and upgradable architecture.

For any given segment, and the set of functions available for its vehicles, the PCU is then tailored to suit the needs of the functions that are to be integrated. This architecture also allows extensibility of the functions by software installation, or the addition of a specific CSS, depending on the extension. This opens the way to an *à la carte* integration of features, usually only accessible to high-end vehicles, even on low-end architectures. Moreover, CSS daughterboards enable hardware maintainability and upgradability at small granularity.

Multiple hardware SoCs have been benchmarked in order to evaluate their performance and the potential to mutualize the use of each SoC across multiple application types, such as image processing, deep neural networks, mathematical optimizations and audio processing. Among potential SoCs, we considered many different computing structures, including embedded microprocessors, embedded GPUs, embedded DSPs, FPGAs and multi- and many-core architectures.

Multiple criteria were used to evaluate the pros and cons of each SoC type, including availability, bring-up time, peak performances, versatility, criticality management, multi-application support, ease of programmability and energy efficiency. Figure 5.3 summarizes our analysis. Deeper insight has also been performed to qualify the adequacy of the different processors to a large set of computing patterns.

Computing Platform	Bring-up	Peak Performances	Versatility	Criticality Management	Multi-Appli	Programmability
Infotainment NXP RENESAS	-	+	+	++	+	++
ADAS/AD SoC NXP RENESAS	-	+	+	+	+	+
DSP TEXAS INSTRUMENTS	-	+	-	-	+	+
Data crunching NVIDIA	++	++	--	--	-	++
FPGA platforms Microsemi XILINX	+	+	+	+	+	-
Many-core KALRAY	-	-	-	+	++	-

Figure 5.3. *Synthetic result of the hardware benchmark*

This approach allows us to define the hardware architecture of the FACE PCU motherboard and its daughterboards in order to serve, in the most efficient way, all the automotive segments.

5.2.2.2. *Modularity and heterogeneity at the software level*

At the software level, the PCU hosts the software functions that run on today's vehicles, plus the advanced driver assistance systems, autonomous driving functions, as well as connectivity and multimedia applications. This diversity presents the challenge of handling the multiple sources of software application heterogeneity, encompassing performance, real-time, safety, security and life cycle.

In addition, the FACE platform hosts together software appliances provided by numerous entities, from corporate software development teams to Tier-1 automotive and Tier-N partner software providers. Consequently, a high-granularity management scheme allows all these software appliances to be developed, validated, updated and deployed independently from each other.

Previous work, especially Becker *et al.* (2015), already demonstrated that embedded virtualization can provide space and time separation between mixed-critical software payloads, ensuring safety up to fail-operational mode in dynamic scenarios. The FACE software platform follows the same principle, going one step further by using the hypervisor to protect software partitions of heterogeneous security, integrity or confidentiality level.

The service-oriented communication architectures (SoA) have been considered for a long time as a suitable paradigm to answer the integration needs of software components in future automotive use cases (Bocchi *et al.* 2008; Cebotari and Kugele 2019). As such, the FACE software platform relies on SoA communications to enable the exchange of data between the components distributed over the PIUs and the PCU, being agnostic of the application programming interfaces (APIs) exposed to the different software suppliers by SoA middlewares (e.g. in the FACE project, the autosar adaptive and the ROS communication middlewares, both exposing different APIs, were integrated).

The problems of indeterminism in SoA communication have been identified in previous work as a problem to generalize the approach to safety-critical components, and the proposed answer is based on monitoring the dataflow between components in the SoA middleware (Frtunikj *et al.* 2014). The FACE software platform embeds a lightweight monitoring layer, seamlessly monitoring the communication behavior of the applications and applying corrective measures when necessary to preserve a deterministic communication behavior. These monitors are automatically generated by the tools described in the next section. In parallel to our work, recent research follows a similar approach (Menard *et al.* 2020). The main distinction in our approach is that we do not need to tag messages to enforce determinism; monitoring is done independently by each software component without impacting any other.

5.2.3. *Tools for specification, configuration and integration*

The third pillar of the FACE project is its software development tooling. Delivering a powerful digital infrastructure, including both hardware and software, is fundamental to succeed in the transition to future car computing architectures, but it is not sufficient. Indeed, such digital infrastructures need specific support to be efficiently exploited and correctly integrated.

Relying on model-based system engineering (MBSE) becomes the only answer to the complexity implied by these challenges. Existing approaches often only rely on denotational models that lack the detailed semantics and property analysis of formal models (e.g. Obergfell *et al.* 2019b). In the context of the automotive industry, some attempts to use formal models in the design process hit the wall of expressiveness, being limited to periodic real-time behavior, and are applicable in practice only to the critical subset of the many functions to integrate (Ernst *et al.* 2018). In addition, formal modeling is in general tightly coupled to the implementation and lacks tools, giving a suitable abstraction level for MBSE.

These drawbacks lead to many difficulties in covering the whole design process. The FACE approach overcomes these limitations by introducing a new tool, the FACE

Tool Suite, relying on a model-based vertical solution, focusing on the emerging role of the software integrator for OEMs (Obergfell *et al.* 2019a). Compared to design tools based on architecture description languages (ADL), such as AADL and EAST-ADL (Mallet *et al.* 2009), FACE Tool Suite uses a formal model that can be integrated in these ADLs.

In FACE Tool Suite, the specification of the functions and their interactions is edited using SysML (OMG 2019), in abstraction of the implementation of the functions themselves. The SysML models in FACE Tool Suite include annotations compatible with the formal model polygraph (Dubrulle *et al.* 2019a,b), covering a wide range of deterministic, real-time, data-driven and dynamic behaviors, well suited to the analysis of complex cyber–physical systems (CPS). The formal execution semantic of polygraph enables the verification of fundamental properties of the system, before even considering its implementation, avoiding costly specification bugs. In addition, this semantic allows a detailed behavior prediction and is used as a common reference in all the other modules of the Tool Suite.

To assist the system designers in an efficient and cost-effective configuration of the modular FACE platform (selection of CSS, VMs and allocation of applications), FACE Tool Suite includes performance prediction and scheduling analysis. For a given configuration, the information on the expected behavior is used at runtime, following a models@run.time philosophy (Blair *et al.* 2009). The configuration of many embedded software components can thus be automatically generated for their target platform (ARXML, etc.). In addition, embedded software and hardware monitors are generated to check at runtime that the functions behave as formally specified, and to apply recovery actions when they do not. This is essential in the approach, since it facilitates the integration of low-criticality software components, developed with less stringent requirements by third parties. These components are expected to behave correctly most of the time but may encounter occasional failures or aberrant behavior. The deterministic and correct-by-construction monitors ensure that their deviations are detected and isolated.

Finally, an automated test infrastructure allows us to validate individual software components by unit testing, and software sub-systems by integration testing, all relying on formal verification methods (Bannour *et al.* 2012; Kirchner *et al.* 2015), automatically configured from the formal polygraph specification.

5.3. Results

All FACE project foundations have been prototyped and integrated into a complete solution. As pre-industrial prototypes, design-to-cost was not the primary concern, and hardware prototypes have been slightly oversized to enable instrumentation and thorough characterization of both hardware and software components.

5.3.1. *Hardware platform*

The FACE electronic architecture relies on the three cornerstones depicted below. The physical computing unit (PCU) is the vehicle main computer, which hosts most software features of the car. It processes data coming from sensors, collected by physical interface units (PIUs) that interface the legacy world with the FACE solution. PIUs communicate with PCU over an Ethernet time-sensitive network (TSN).

5.3.1.1. *Physical computing unit*

The PCU is the core of the FACE infrastructure. To support the whole range of vehicles in the Renault–Nissan–Mitsubishi Alliance, and the diversity of products within vehicle segments, a modular approach is implemented. It allows us to scale PCU performance with respect to vehicle feature performance needs.

The backbone of the PCU is a dual-sided PCB whose high-level overview is given in Figure 5.4. The front side, known as the motherboard, hosts all the basic components that allow us to satisfy the requirements common to all vehicle ranges and options. The back side gives us access to four sets of connectors, making it possible to populate the PCU with daughterboards.

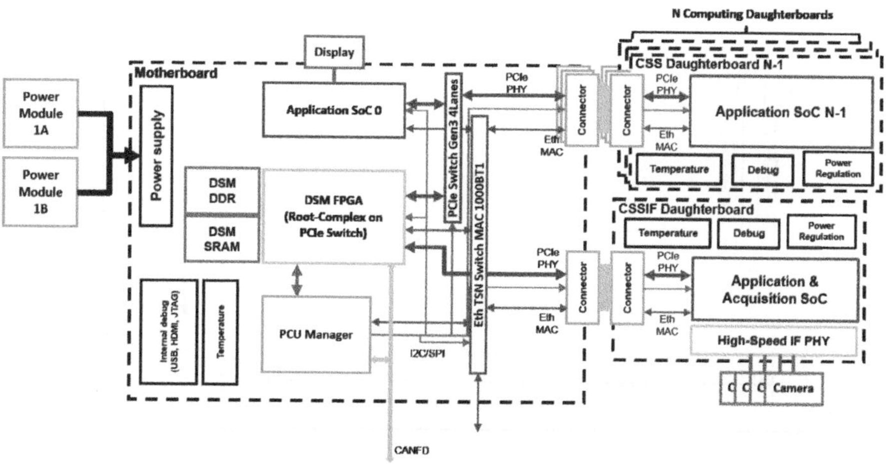

Figure 5.4. *Overview of the FACE PCU structure*

The motherboard includes a high-performance general-purpose system-in-package leveraging on an RCar H3 SoC from Renesas. This automotive-grade (AECQ-100) SoC runs at 1.6 GHz and is built around the ARM® Cortex®-A57/A53 and Cortex®-R7 cluster. It achieves up to 40,000 DMIPS and embeds a couple of accelerators for graphics and multimedia processing. This application SoC is managed, as the rest of the platform, by the PCU manager.

The PCU manager relies on a Renesas RH850-F1H premium processor. This safety-oriented (ISO2626-2 ASIL-D capable) micro-controller runs at 120 MHz and communicates with all the PCU resources through dedicated interrupts and I2C links. For higher availability, the PCU is powered by two redundant 12V power supplies. The PCU manager is the first component that starts on the PCU. It controls the eight power domains and boots the different computing sub-systems (CSS), configures the communication infrastructure, loads the configuration and controls their execution. The PCU manager then acts as a safety supervisor of PCU and can shut down or reset malfunctioning CSS at any time during execution.

Other CSS are integrated as daughterboards. These boards have a fixed form factor (see Figure 5.5) and are plugged into the motherboard with two dedicated Hirose FX23 connectors. Each connector has four 3A/12V power pins but differ in the signals they transport. The 40-pin connector transports signals enabling four full-duplex PCIe lanes. To support high-frequency PCIe gen3 communications, the signal integrity is preserved with the interleaving of two ground signals between each differential pair. The second connector has 60 pins shared between Gbps Ethernet and controls signals (I2C, GPIO, status, power management, temperature warning, interrupts).

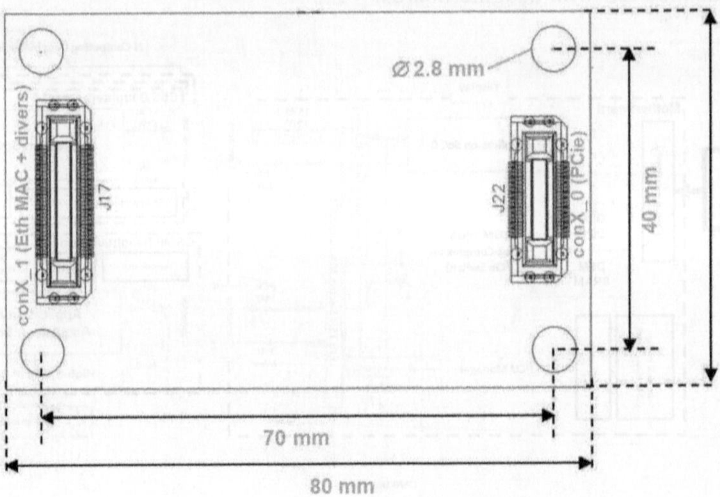

Figure 5.5. *Daughterboard physical form factor*

The motherboard offers a high-bandwidth communication backbone between all CSS, and to the external vehicle network. Cumulated bandwidth exceeds 200 Gbps thanks to the combination of two communication paths. An internal automotive Ethernet network (TSN) manages internal PCU communications with a theoretical bandwidth of 1 gigabit per second (Gbps), while PCIe Gen3/4-lanes network manage simultaneously up to six 32 Gbps links for large payloads. These networks are routed through a set of switches, NXP's SJA1101 for Ethernet TSN and Broadcom PEX8725 for PCI-express.

A dedicated hardware multi-protocol gateway, called the data-sharing manager core, is implemented in a Xilinx Zynq Ultrascale+ ZU5EV. The DSM core bridges the PCIe with both internal and external Eth TSN networks. The DSM core schedules communications according to an off-line determined schedule, to eliminate interferences due to bus sharing. The DSM core detects communication and timing faults on the PCIe bus, increasing the reliability of the data exchange on this medium. It also prevents access rights infringements by construction, limiting access to sensitive data only to allowed connected CSS. This component thus allows safe and secure sharing of data from all over the vehicle by all adequately privileged applications on the PCU, with a high benefit for fusion algorithms that require both complementary and redundant usage of sensor data (Kaempchen and Dietmayer 2003). The schedule and access right management are generated from the formal model by the FACE Tool Suite, contributing to both safety and security partitioning between computing domains of heterogeneous safety or the security level.

The current daughterboard portfolio includes two types of programmable CSS. The first one, designed by Renesas, leverages on the RCar-M3 processor. It allows us to extend the general-purpose computing processing capabilities with an ARM® Cortex®-A57/A53 quad core solution, running at 1.6 GHz. The second board, designed by Kalray, takes benefit of their Bostan many-core processor. The 288 cores of the component and its massively parallel processing capabilities make it relevant for neural networks or computer vision, as well as real-time mixed-critical payloads.

A specific daughterboard completes the portfolio to manage raw data from high-bandwidth sensors, such as a high-definition camera or LIDAR, directly at the PCU level. The board includes two sets of four FAKRA-Mini connectors, using FPDLink III or GMSL2 serdes. Raw data are transmitted to an automotive-grade Xilinx Zynq Ultrascale+ ZU5EV, through four MIPI-CSI2 signals. Low-latency processing capabilities of the device allow us to pre-process the raw data on the fly, completing some data fusion or compressing before transmitting the data to other computing sub-systems, as depicted in Piriou *et al.* (2020).

5.3.1.2. *Physical interface unit*

The physical interface unit (PIU) interfaces the legacy world with the FACE platform. The legacy world is mostly composed of the very large set of sensors and actuators distributed all along the car, and includes a specific ECU (electronic control unit) that will not be centralized for reactivity or business concerns. To manage the diversity of sensors, actuators and ECU, the PIU supports a very large set of communication protocols such as CAN, CAN FD, LIN, SPI and analog PWM interfaces.

The first duty of the PIU is thus to behave as a smart gateway to manage the diversity of sensors/actuators and aggregate data to be exchanged with the PCU through the vehicle Eth/TSN network. A communication scheme between PCU and PIUs is determined offline for each execution mode, to optimize trade-off between bandwidth and latency, and guarantee data transport in known and bounded time. Up to 180 heterogeneous inputs/output are accessible through four connectors. Such a huge quantity of IOs cannot be handled by a programmable processor in a short time. As depicted in Figure 5.6, we use a Xilinx Ultrascale+ ZCU5EV to handle these IOs in a time-predictable, low-latency manner. This FPGA, called DSM Node, packs and unpacks IO data payloads into/from TSN Ethernet frames according to the communication schedule generated from FACE Tool Suite. The functionalities it offers to the connected legacy equipment are similar to the one offered by the DSM core to the CSS, enforcing a coherence in the schedule between legacy media (CAN, LIN, etc.), automatically generated in FACE Tool Suite from the information extracted from the formal model.

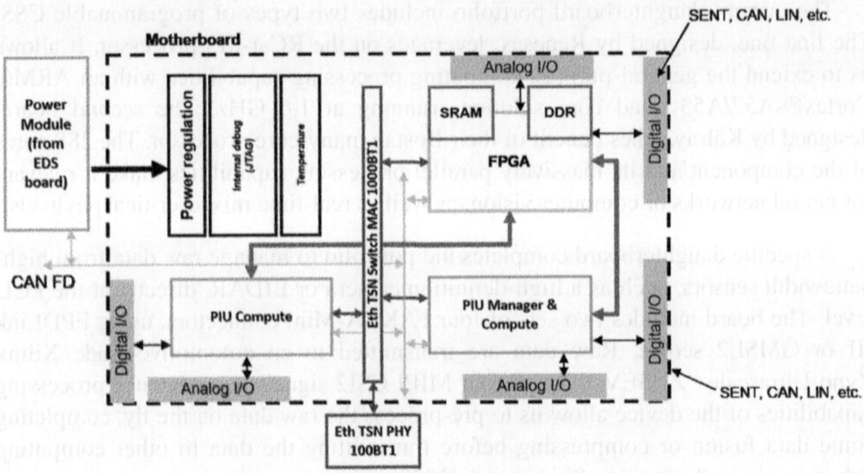

Figure 5.6. *Overview of the FACE PIU structure*

In addition to the FPGA, the PIU integrates an RH850 micro-controller. It oversees the configuration management of the smart gateway and overall board health monitoring. For some reflex features that may need response times below 5 ms, we also integrate a second micro-controller, running at 120 MHz to execute safety-oriented control operations.

5.3.1.3. *Network ETH/TSN*

The FACE architecture intrinsically requires that multiple types of applicative traffic be transported over a common physical communication link between PCU and PIUs. The variety of intra-vehicular automotive applications makes this challenge especially complex. Indeed, instrumentation and control applications on the one hand typically require predictably low latency while accommodating minimal bandwidth. On the other hand, camera- or LIDAR-generated data require large bandwidth while supporting occasional (though bounded) delivery delay, provided that each sample be adequately timestamped.

These two examples only (over a dozen of relevant applicative traffic types) explain why specific mechanisms are required for managing the intra-vehicular network. They also explain why simply over-dimensioning the physical links (say, optical fiber with legacy communication protocols) would be a false good idea: in most situations, all applicative types of traffic would flow seamlessly over these links – but at some points, a succession of events would lead to excessively long link unavailability, which would in turn lead to unacceptable applicative latency.

The answer to the need for carrying multiple applicative payloads with multiple respective quality of service constraints over a single physical wired medium is the reliance on the time-sensitive networking (TSN) paradigm. TSN represents a set of standards that define how the Ethernet can be adapted to transport data while affording them with deterministic properties. In a nutshell, TSN can guarantee the availability of a given bandwidth dedicated to a class of applications, achieve time-bound delivery or even ensure controlled variance over the delivery latency. Throughput, latency and jitter are therefore fully controllable by means of the protocols defined in the framework of 802.1 TSN. In turn, these protocols achieve specific functions needed for these higher-level requirements such as node synchronization (802.1AS-rev), credit-based shaping (802.1Qav) and time-aware shaping (802.1Qbv).

The very rich toolbox provided by the set of Ethernet TSN standards is more than enough to answer the applicative QoS requirements. Nevertheless, the proper choice and configuration of the different TSN protocols it is made up of require a specific planning step. This planning fulfills the twofold objective of 1) determining the configuration parameters of every required TSN protocol at every network node and 2) ensuring that the chosen set of TSN protocols, configured in the chosen manner, will always be sufficient to prevent packet loss or out-of-delay packet delivery. Therefore, we relied on an RTaW-pegase advanced simulator that gave us the assurance that the proposed set of TSN standards would meet its goals even in worst-case situations. Figure 5.7 shows an example of a FACE platform setup relying on a TSN network.

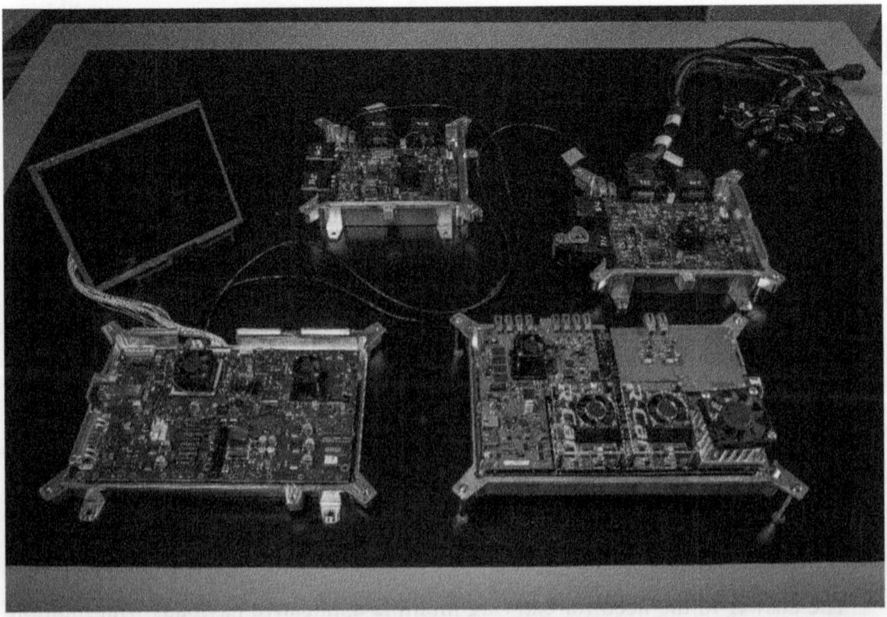

Figure 5.7. *Example of hardware setup of the FACE platform. PCU front and back sides are on the bottom of the photo. PCU is connected through 100Base-T1 Ethernet TSN to two PIUs. Harness to connect heterogeneous sensors and actuators is in the right-top corner of the photo*

5.3.2. *FACE SW architecture*

5.3.2.1. *Layered stack structure*

The software architecture, or software stack, deployed onto the FACE hardware infrastructure is decomposed into several layers. Each layer has a specific responsibility and may be provided and/or configured by a specific software provider. Main layers are the board support package, the hypervisor, the operating system, the middleware and generic services and the application layer.

5.3.2.1.1. Board support package

The board support package provides the set of software drivers that enable low-level interaction with the SoC-level and board-level peripherals. FACE BSP is built from SoC vendors' basic software, patched and completed to accommodate the PCU's specific configuration.

5.3.2.1.2. Hypervisor

The hypervisor is deployed on the applicative SoC and general-purpose CSS. It can host several "guest" virtual machines, each with its own OS, and manages guest access to hardware resources (CPU time, memory, peripherals access) or shared services (e.g. virtual network bridge). This hypervisor is bound to the SoC BSP.

After a methodological selection process described in Hamelin *et al.* (2020), the PikeOS hypervisor from Sysgo was chosen as the reference hypervisor. After characterization, it appeared that the hypervisor by itself does not represent a significant overhead on application computing time, response latency or memory utilization. We showed however that virtualized guest applications can interfere with each other through shared resources (e.g. shared caches, memory buses or virtualized services), hence causes significant variation or degradation of the execution performance of application code. As a mitigation, we reduce the use of shared resources between time- or safety-critical appliances and non-critical ones. For instance, on the RCar-H3 SoC, the Cortex-A57 cluster is dedicated to memory-hungry full-virtualized OSs (e.g. Linux), while the Cortex-A53 cluster runs real-time applications from its separate L1 cache, and the Cortex-R7 cluster (in lock-step configuration) runs high-criticality safety monitors from its tightly coupled memory.

In the FACE software architecture, the hypervisor ensures space and time separation between heterogeneous virtual machines. VMs can thus be managed by independent software application suppliers, developed, tested, deployed and upgraded independently of each other – provided that each one is dimensioned according to a worst-case interference scenario.

The vehicle software integrator is responsible for configuring the hypervisor with access rights and quota to ensure safe and secure sharing of hardware resources or virtual services between virtual machines. This task can be assisted by the static analysis of the predicted interactions in the formal model of FACE Tool Suite.

5.3.2.1.3. Operating system

Each VM then runs its own operating system, responsible for scheduling execution threads on hypervisor-monitored resources. A variety of OSs can be used inside guest VMs, for example, Linux-based Android for infotainment applications, real-time Linux distribution for some control applications or even AUTOSAR Classic Platform OS to ease the integration of legacy applications. We chose a real-time Linux distribution (Yocto Linux with PREEMPT-RT patches) in our evaluation.

The OS requires direct interaction with some hardware IPs, and is thus bound to the BSP's drivers. In some cases, a set of paravirtualized drivers are provided by the hypervisor editor, to enable interaction between the OS and hypervisor services. For instance, Sysgo provides a virtual network bridge service driver for Linux, which enables several guests to communicate over a single hardware Ethernet interface.

5.3.2.1.4. Middleware

Over the OS, the middleware is a framework of generic utility software functions, specifically supporting service-oriented application. Within FACE, we designed the software infrastructure around both the AUTOSAR Adaptive Platform and the ROS robotics framework, to illustrate flexibility offered by an SOA framework over the heterogeneous hardware topology.

The AAP framework, especially the AAP-1903 release, provides a set of standard APIs and service interfaces that implement many generic utility software functions, as depicted in Figure 5.8. Of particular interest for the FACE platform is the ARA communication management cluster (ARA Com). This framework provides transparent serialization and adaptive routing of messages between software components (SWCs), thus supporting a service-oriented architecture design of application software.

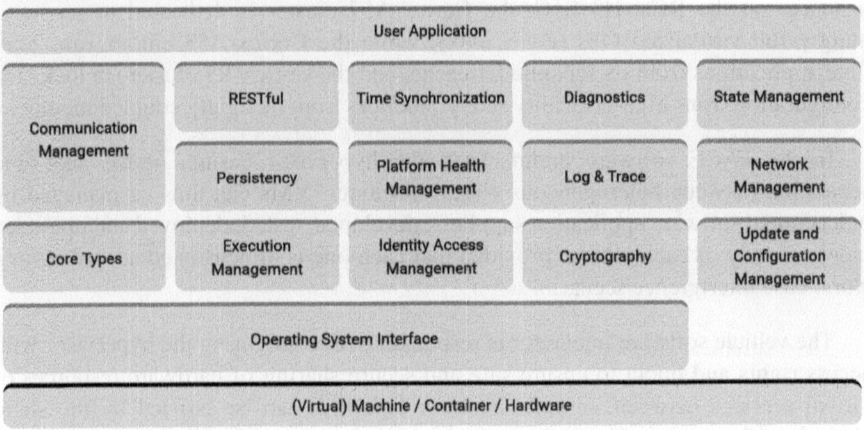

Figure 5.8. *Illustration of AUTOSAR adaptive platform software architecture*

In the SOA paradigm, a software application is composed of a set of SWCs, interacting through their service interfaces. Real-time SWCs send and receive events in a timely manner so that the composition of all SWCs respects end-to-end timing constraints (e.g. sensor to actuator latency). In order to support a contractualization of application SWC interfaces, including in timing, we add a monitoring layer between the middleware and the application.

We provide an event timing model, as a realization of the formal model supported in the FACE Tool Suite modeling platform. We have developed a specific software monitoring layer over AAP, called the data sharing management monitor (DSMMon). This allows us to automatically configure the ARA platform health management (ARA PHM) and ARA Com services, from timing and communication constraints extracted by the analysis of the formal model in FACE Tool Suite.

The DSMMon monitors communication events, i.e. calls to the ARA Com API, for each SWC selected for online monitoring in the FACE Tool Suite. If an event timing is correct, then the event is transparently forwarded to the ARA Com framework. However, if an SWC suffers a timing failure or makes an unexpected call to communicate, the DSMMon intercepts faulty events and triggers a recovery mode or fault handler so that the timing fault is not propagated to other SWCs.

Several recovery actions are available, configured in FACE Tool Suite per ARA communication interface. If an SWC attempts to send too many or too few event samples (over-/under-production of events), the DSMMon may restart the faulty SWC, switch to degraded mode, if the SWC is critical, or silently discard excessive samples in less critical cases. If an expected sample arrives late (or never arrives), the DSMMon may provide a default value or signal the fault to the consumer SWC. Similarly, in case of input events over-/under-consumption, the DSMMon may discard excessive read requests, drop outdated incoming samples or trigger fault handlers. DSMMon uses the ARA PHM services for fault detection and management.

5.3.2.1.5. The application

The application software is designed and implemented as a set of AAP SWC or ROS nodes, interacting through SOA services. The applications may be provided by several Tier-N software suppliers.

If the application uses standard AAP services, and is correctly modeled with its timing specification in FACE Tool Suite, it shall run correctly over the middleware and DSMMon layer. The DSMMon is transparent to the application code. The measured latency of this monitoring stage is of the order of a few hundred microseconds. This contract-based approach allows the OEM to deploy tier applications onto any compatible CSS on the PCU: if the system is schedulable, then execution is functionally deterministic.

5.3.2.2. *Virtual prototyping framework*

Due to the complexity of SW stack, the amount of hardware-dependent software and the simultaneous development of hardware and software, testability was a key concern in the project. The software design team leveraged on a specific test environment, a digital twin of the real platform prototype.

The built virtual platform (FACEVP) pursues the below objectives that result in the definition of a set of high-level requirements:

– to have a pre-validated firmware available for the bring-up of the boards;

– to perform validation of IP models within a representative system environment;

– to provide software developers with a test environment prior to boards' availability.

First, the FACEVP executes the same binary code as the final target. This allows one to maintain only one load build. The virtual platform also catches most of the hardware-dependent issues. Second, the FACEVP interconnects with FPGA vendor simulators to validate in advance IPs. Third, the FACEVP is flexible enough to be representative of one board (PCU motherboard), a module (PCU, PIU) or various FACE platform configurations (one PCU connected to one or several PIUs). Fourth, the FACEVP accepts the use of regular debug tools and completes their investigation capabilities with hardware behavioral inspection capability. Finally, the FACEVP is fast enough to show its users an execution time in the same order of magnitude than on the real HW.

The FACEVP was built using the CEA-owned VPSIM framework (Charif *et al.* 2019). This simulation environment allows us to describe and build a virtual platform by integrating models from various origins onto a system C kernel. Looking at the main building blocks, the FACEVP is a combination of QEmu open-source models (for ARM-based CPUs), an OVP model (for the RH850 processor), a homemade PCIe model and a VirtIO-based Ethernet network model. The validation of IPs within this platform relies on Xilinx ModelSIM connected to the FACEVP through VPSIM's co-simulation interface (Wicaksana *et al.* 2019).

The FACEVP uses several modeling abstractions and therefore is not cycle-accurate; however, through comparing various measurements performed on the virtual prototype of the RCAR-H3 and the real board, we were able to show that the FACEVP could capture some essential characteristics of software applications, such as cache misses and relative timings, especially in the case of long-running applications. In this case, the FACEVP measured performance is a useful approximation (20% error) of the actual target performances.

5.3.3. *FACE Tool Suite*

Within FACE Tool Suite, we have implemented a methodology conforming to "ISO 26262 Road Vehicles – Functional Safety" (*ISO 26262-1:2018* 2018). The following sections describe the different features of FACE Tool Suite, part of its methodology.

5.3.3.1. *The functional architecture modeling feature*

A system is modeled in the formal model polygraph. Each software component is modeled as an actor, an abstract entity representing the behavior of the component. The actors can communicate through unidirectional FIFO channels. This model captures the communication dependencies and service interfaces in an SoA. The publication of events by a component through a service interface and the subscription to these events by another component are indeed a direct transposition of an actor producing data tokens to a channel and another actor consuming them. The same stands for the sending of parameters in a request for a method call and so on.

To capture the real-time requirements of physical interfaces (sensors/actuators), it is possible to label the corresponding actors with a frequency constraint. The actor is then constrained to send, receive and process data (i.e. fire) at this frequency, in global synchrony with all other actors that also have a frequency constraint. Actors that do not have such a frequency label are not constrained and may fire as soon as enough input data is available to them. An example of a polygraph model, in FACE Tool Suite, is shown in Figure 5.9. Sensor actors are allocated on the PIU, while other actors are allocated on the PCU.

This notion of enough data is provided by rate annotations on the channels, specifying how many samples are produced or consumed per firing. Rational communication rates can be specified to express a resampling of the data stream to adjust it to the firing frequencies. Thanks to that information, coupled to the frequency labels, the communications are determinate, and the production/consumption time of a sample can be inferred. This is particularly useful for data fusion algorithms. More details on this aspect can be found in Dubrulle *et al.* (2019b).

Polygraph annotations also include the expression of execution modes. This notion allows the expression of data-dependent reconfigurations of the system. An actor may declare that it decides of different modes. For example, the component reading the current speed from an odometer sensor may be modeled as a frequency-constrained actor declaring three modes corresponding to the speed ranges 0–30, 30–50 and above 50 km/h. Based on such modes, any actor may declare that its behavior is modified by the current mode declared by another actor. Continuing the odometer example, a component in charge of crash prediction with different execution times (potentially 0, i.e. deactivated) depending on the current speed may be modeled as an actor depending on the modes produced by the odometer and define its alternative behaviors accordingly. This allows the capture of deterministic dynamic transitions between operational modes in the system. More details on this aspect can be found in Dubrulle *et al.* (2019a).

Some platform-specific properties may be annotated. An actor has a worst-case execution time (WCET) and a best-case execution time (BCET). Actor activation may have jitter. An actor is allocated on a processor, and it may have a fixed priority for such scheduling policies.

5.3.3.2. *About model validation features*

In addition to the modeling feature, FACE Tool Suite provides its users with formal validation of polygraph properties, simulation and schedulability analysis.

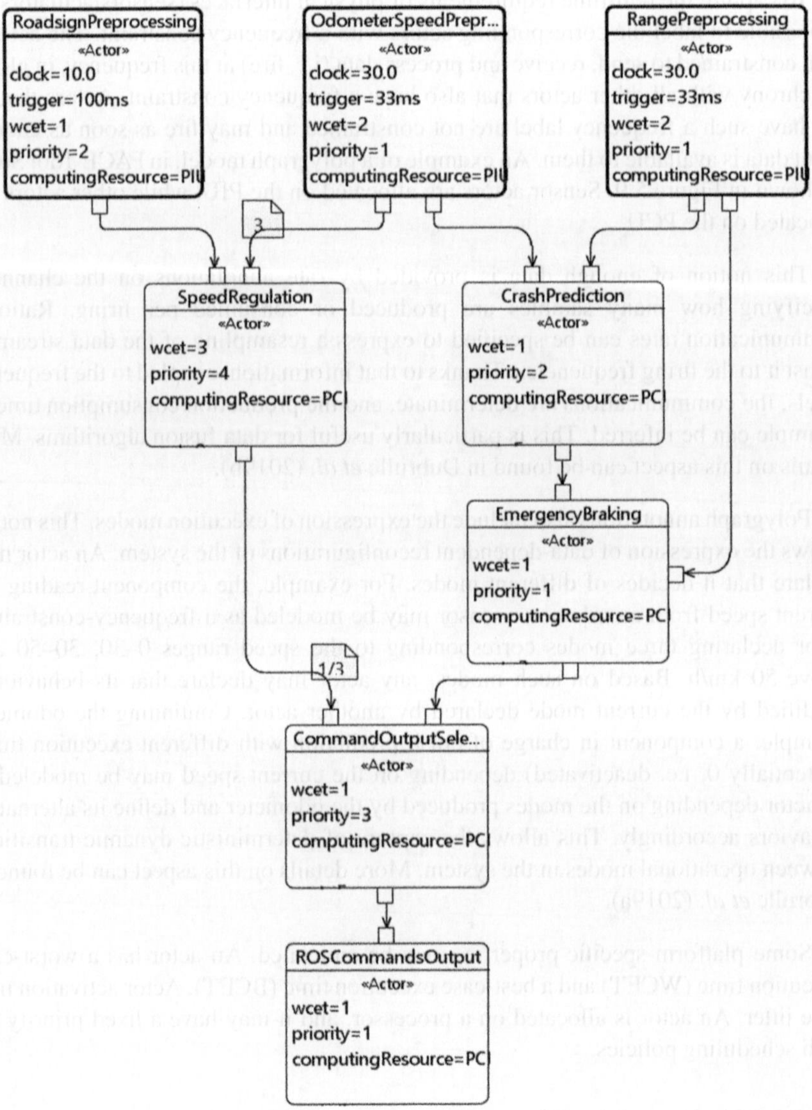

Figure 5.9. *ADAS polygraph model*

Two classic SDF properties must be validated for a polygraph model: consistency and liveness. Consistency guarantees bounded-memory cyclic behaviors (in other words, it is possible to implement the communications without losing data samples), and liveness guarantees the absence of deadlocks. The validation of both properties is based on polygraph theory that extends SDF theory (Dubrulle *et al.* 2019b).

Another property, coherence, is specific to the execution modes introduced by polygraph. It guarantees that dynamic reconfigurations in the system do not lead to missing or invalid input data. The validation features also check this property and detect undesirable practical situations such as data produced in a previous configuration remaining in buffers of a component, resulting in incoherent computation on its next execution. The validation of this property is achieved by symbolic execution on a minimal period of operation of the modeled system (Dubrulle *et al.* 2019a).

A simulator has been implemented for polygraph, to allow the system designers to check that their model is not only correct (properties mentioned above) but also that it is the right model, i.e. it models the behavior they expected. To do this, it is possible to place breakpoints on model elements, such as channels. This allows us to suspend the simulation and check the state of the model. The model may be animated during simulation.

On the FACE platform, polygraph actors are implemented as tasks scheduled online. For this reason, FACE Tool Suite provides schedulability analysis. Like Singh *et al.* (2017), we therefore apply classic real-time scheduling theory.

Our schedulability analysis method consists of transforming a polygraph model into a transactional task model (Palencia and Harbour 1999): periodic/sporadic tasks (Liu and Layland 1973), offset, jitter and precedence dependencies, allocated onto a partitioned multi-processor platform. As in Li *et al.* (2017), we propose to automatically model jobs of actors, as sporadic tasks, with different precedence dependencies due to communication rates.

One of the advantages of a polygraph model is the existence of untimed actors, who do not have a frequency. In our analysis, we infer an average frequency for such actors, based on the results of the consistency property verification (Dubrulle *et al.* 2019b). For example, the average frequency of command output selection is 30 Hz. This frequency is an average and does not require a strict periodic behavior for the corresponding task. This allows further refinement of the transformed model, to optimize schedules by pipelining the execution of such tasks. This approach avoids the performance cost of scheduling them as strictly periodic tasks. We perform schedulability analysis on the task model with the compositional performance analysis approach (Schlatow and Ernst 2016; Diemer *et al.* 2019), because it is well adapted to many graphs of tasks.

5.3.3.3. *About the deployment feature*

In order to deploy the functional architecture model, the user must transform it into a component-based model. The software components have implementation details such as operations and attributes. The component-based model also refines the data types sent between software components.

From the software components model, it is possible to generate the equivalent descriptions of AUTOSAR software components as ARXML files. ARA Com interface classes may also be generated for each software component.

5.3.3.4. *About the integration validation feature*

Within the methodology, we also generate configuration data for the online software and hardware monitors introduced in the previous sections. The monitors also produce execution traces that are used by FACE Tool Suite to automatically validate the integration's conformity to the polygraph model (Bannour *et al.* 2012; Dubrulle *et al.* 2019b).

The trace can also be replayed in the model to explain the cause of a problem. The same debug environment used for model validation is also used for trace replay. With such a feature, it is possible, for example, to identify data lost in the network on non-critical streams scheduled in best effort.

5.4. Use case

The methodology, tooling and hardware/software architecture presented in section 10.5 is demonstrated and characterized by implementing an adaptive braking system use case.

5.4.1. *Adaptive braking system*

The FACE instrumented prototype supports an ADAS system, as depicted below. A desktop PC equipped with the ROS middleware plays pre-recorded video and odometer data from the KITTI datasets (Geiger *et al.* 2013), and a PIU sends real-time LIDAR data to the PCU, where computation is distributed among several computing sub-systems.

Seven actors are implemented as software components (SWCs) with hard real-time constraints and deployed on a platform configured as depicted in Figure 5.10. They are developed for the AUTOSAR adaptive platform environment, inside a real-time Linux virtual machine run by the PikeOS hypervisor on the RCar H3 application SoC of the PCU main board. These SWCs implement the road sign, odometer speed and range preprocessing, speed regulation, crash prediction, emergency braking and command output selection nodes modeled in section 5.3.3. They are monitored by the DSMMon software monitoring layer. Each SWC runs in less than 2 ms, the DSMMon represents an overhead lower than 100 microns, resulting in a CPU utilization under 6% each.

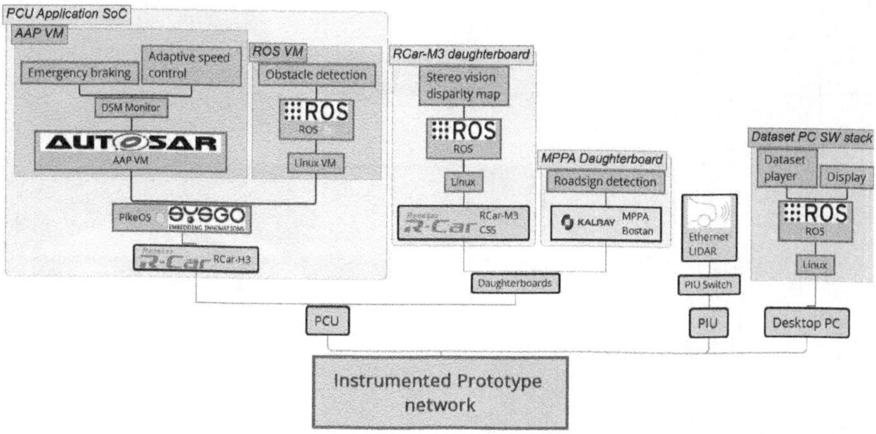

Figure 5.10. *The FACE instrumented prototype setup*

Moreover, these AAP SWCs interact with several soft-real-time actors implemented as ROS nodes through an AAP/ROS bridge, where ROS nodes either run within a Linux virtual machine hosted by the PikeOS hypervisor on the application SoC or on a native Linux deployed on an RCar-M3 computing sub-system. The MPPA Bostan daughterboard hosts the speed limit road sign detection algorithm.

These software nodes illustrate the diversity of computation, from video image processing with both classical computer vision and neural network algorithms (speed limit road sign recognition, pedestrian detection), occupancy grid mapping, while the AAP software components perform real-time control. The rendering of the result of these various functions is illustrated in Figure 5.11.

In this use case scenario, we demonstrate that the hardware and software infrastructure of the PCU supports a controlled integration of heterogeneous software applications, deployed onto several virtual machines and several computing sub-systems on the PCU, interacting in real-time with external sensors and actuators.

This architecture supports dynamic reconfigurations. First, depending on vehicle speed, the runtime switches between several off-line validated execution modes, and activates or deactivates several features, for example detection of pedestrians with disabilities and lane departure warnings when the speed is above 50 km/h. Second, a sensor or computing sub-system can be switched on/off at runtime: the runtime environment detects function availability status and updates data routes accordingly.

Moreover, if a fault is injected into a software component (e.g. a job expends its time budget or does not deliver its output data on time), then DSMMon detects the temporal fault and prevents its propagation to other components.

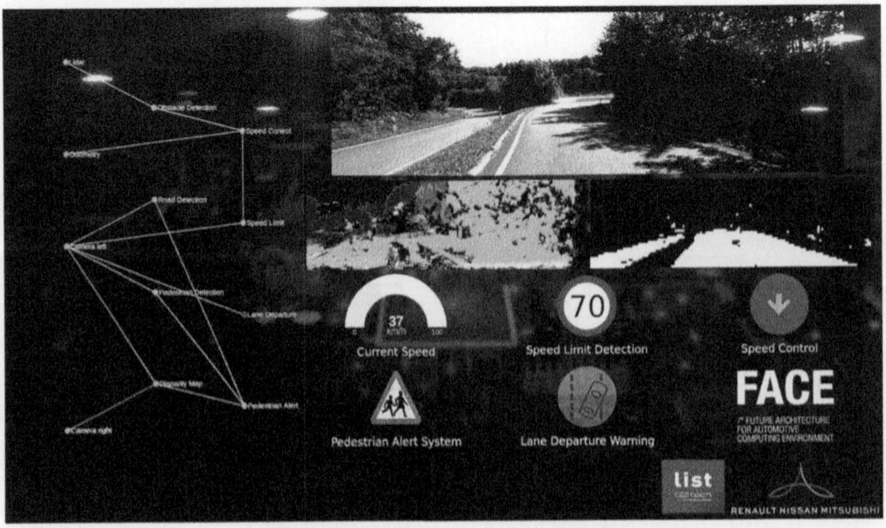

Figure 5.11. *Use case interface*

5.5. Conclusion

Centralization is a trend that affects the whole automotive industry. It solves a couple of concerns born from a long history of incremental innovations in electrical and electronic architectures. If most centralization initiatives are done at domain levels, we show in this chapter the relevance of addressing the centralization at the vehicle scale. This step forward in embedded electronics also comes with significant changes in the automotive supply chain, as vehicle computer can now be open to a very large community of software application developers.

Through a system-level approach, FACE provides means to vehicle manufacturers and suppliers to leverage the value of centralized computers by managing features integration during the whole vehicle life cycle. We highlight through the FACE platform deployment example that we address the five key concerns during features integration:

– safety and real time are addressed all along the SW integration process through an innovative model of computation and ISO26262 tools;

– multi-sourcing management is enabled by the conformity checking that is shown to be of linear complexity according to the amount of feature providers;

– execution framework diversity is highlighted with the FACE platform that supports five different CPUs and three operating systems;

– HW sizing is almost reduced by half thanks to the support of dynamic reconfiguration and asynchronous communications support;

– products range can be very large thanks to the modular architecture support enabling the addition of multiple high-end daughterboards increasing computation power from 40kDMIPS to 400kDMIPS.

Behind these technical results, the FACE project illustrates the impact of joint hardware, software and the development of tools. The cross-fertilization of multiple skills involved in the program made it possible to promote real disruptions and to make them acceptable even for a conservative industry such as the automotive industry. Next steps consist of moving towards industrialization through a design-to-cost optimization and the application of the FACE foundations to other mobility industries (trucks, defense, logistics, etc.).

5.6. References

Bannour, B., Escobedo, J.P., Gaston, C., and Le Gall, P. (2012). Off-line test case generation for timed symbolic model-based conformance testing. In *Testing Software and Systems*, Nielsen, B. and Weise, C. (eds). Springer, Berlin, Heidelberg, pp. 119–135.

Becker, K., Frtunikj, J., Felser, M., Fiege, L., Buckl, C., Rothbauer, S., Zhang, L., and Klein, C. (2015). RACE RTE: A runtime environment for robust fault-tolerant vehicle functions. *CARS 2015 – Critical Automotive Applications: Robustness & Safety*, Paris, France [Online]. Available: https://hal.archives-ouvertes.fr/hal-01192987.

Blair, G., Bencomo, N., and France, R.B. (2009). Models@run.time. *Computer*, 42(10), 22–27.

Bocchi, L., Fiadeiro, J., and Lopes, A. (2008). Service-oriented modelling of automotive systems. *32nd Annual IEEE International Computer Software and Applications Conference*, pp. 1059–1064.

Cebotari, V. and Kugele, S. (2019). On the nature of automotive service architectures. *2019 IEEE International Conference on Software Architecture Companion (ICSA-C)*, pp. 53–60.

Charif, A., Busnot, G., Mameesh, R., Sassolas, T., and Ventroux, N. (2019). Fast virtual prototyping for embedded computing systems design and exploration. *Proc. of RAPIDO2019 – Workshop on Rapid Simulation and Performance Evaluation: Methods and Tools*, Valence, Spain, pp. 1–8 [Online]. Available: https://hal.archives-ouvertes.fr/hal-02023805.

Diemer, J., Axer, P., Thiele, D., and Schlatow, J. (2019). pyCPA [Online]. Available: https://pycpa.readthedocs.io/en/latest/index.html.

Dubrulle, P., Gaston, C., Kosmatov, N., and Lapitre, A. (2019a). Dynamic reconfigurations in frequency constrained data flow. In *Integrated Formal Methods*, Ahrendt, W. and Tapia Tarifa, S.L. (eds). Springer International Publishing, pp. 175–193.

Dubrulle, P., Gaston, C., Kosmatov, N., Lapitre, A., and Louise, S. (2019b). A data flow model with frequency arithmetic. In *Fundamental Approaches to Software Engineering*, Hähnle, R. and van der Aalst, W. (eds). Springer International Publishing, pp. 369–385.

Ernst, R., Kuntz, S., Quinton, S., and Simons, M. (2018). The logical execution time paradigm: New perspectives for multicore systems (Dagstuhl Seminar 18092). *Dagstuhl Reports*, 8, 122–149 [Online]. Available: https://hal.inria.fr/hal-01956964.

Frtunikj, J., Rupanov, V., Camek, A., Buckl, C., and Knoll, A. (2014). A safety aware run-time environment for adaptive automotive control systems. *Embedded Real-time Software and Systems*.

Geiger, A., Lenz, P., Stiller, C., and Urtasun, R. (2013). Vision meets robotics: The KITTI dataset. *The International Journal of Robotics Research*, 32, 1231–1237 [Online]. Available: https://doi.org/10.1177/0278364913491297.

Hamelin, E., Ait Hmid, M., Naji, A., and Mouafo-Tchinda, Y. (2020). Selection and evaluation of an embedded hypervisor: Application to an automotive platform. *European Congress of Embedded Real Time Software and Systems*.

ISO 26262-1:2018 (2018). ISO 26262-1:2018 [Online]. Available: http://www.iso.org/cms/render/live/en/sites/isoorg/contents/data/standard/ 06/83/68383.html.

Kaempchen, N. and Dietmayer, K. (2003). Data synchronization strategies for multi-sensor fusion. *Intelligent Transportation Systems*, vol. 85-1, pp. 1–9.

Kirchner, F., Kosmatov, N., Prevosto, V., Signoles, J., and Yakobowski, B. (2015). Frama-C: A software analysis perspective. *Formal Aspects of Computing*, 27(3), 573–609 [Online]. Available: https://doi.org/10.1007/s00165-014-0326-7.

Li, S., Singhoff, F., Rubini, S., and Bourdelles, M. (2017). Scheduling analysis of tasks constrained by TDMA: Application to software radio protocols. *Journal of Systems Architecture*, 76, 58–75 [Online]. Available: http://www.sciencedirect.com/science/article/pii/S1383762116301989.

Liu, C.L. and Layland, J.W. (1973). Scheduling algorithms for multiprogramming in a hard-real-time environment. *Journal of the ACM*, 20(1), 46–61 [Online]. Available: https://doi.org/10.1145/321738.321743.

Mallet, F., Peraldi-Frati, M.-A., and André, C. (2009). Marte CCSL to execute East-ADL timing requirements. *2009 IEEE International Symposium on Object/Component/Service-Oriented Real-Time Distributed Computing*, pp. 249–253.

McKinsey Company (2019). Automotive software and electronics in 2030 [Online]. Available: https://www.mckinsey.com/industries/automotive-and-assembly/our-insights/mapping-the-automotive-software-and-electronics-landscape-through-2030.

Menard, C., Goens, A., Lohstroh, M., and Castrillón, J. (2020). Achieving determinism in adaptive AUTOSAR. *Proceedings of Design Automation and Test in Europe* [to appear], 9–13 March 2020.

Obergfell, P., Kugele, S., Segler, C., Knoll, A., and Sax, E. (2019a). Continuous software engineering of innovative automotive functions: An industrial perspective. *2019 IEEE International Conference on Software Architecture Companion (ICSA-C)*, Hamburg, Germany, pp. 127–128.

Obergfell, P., Sax, E., and Kugele, S. (2019b). Model-based resource analysis and synthesis of service-oriented automotive software architectures. *IEEE/ACM 22nd International Conference on Model Driven Engineering Languages and Systems (MODELS)*, Munich, Germany, 15–20 September.

OMG (2019). OMG System Modeling Language – SysML 1.6 beta, Specification ptc/18-12-08. Technical Report. Available: https://www.omg.org/spec/SysML/1.6/Beta1/PDF.

Palencia, J. and Harbour, M. (1999). Exploiting precedence relations in the schedulability analysis of distributed real-time systems. *Proceedings of the 20th IEEE Real-time Systems Symposium*, pp. 328–339.

Reinhardt, D. and Kucera, M. (2013). Domain controlled architecture – A new approach for large scale software integrated automotive systems. *Pervasive and Embedded Computing and Communication Systems*, pp. 221–226.

Schlatow, J. and Ernst, R. (2016). Response-time analysis for task chains in communicating threads. *2016 IEEE Real-time and Embedded Technology and Applications Symposium (RTAS)*, Vienna, pp. 1–10.

Singh, A., Ekberg, P., and Baruah, S. (2017). Applying real-time scheduling theory to the synchronous data flow model of computation. In *29th Euromicro Conference on Real-time Systems (ECRTS 2017)*, Bertogna, M. (ed.). Schloss Dagstuhl–Leibniz-Zentrum fuer Informatik, Dagstuhl, Germany, pp. 8:1–8:22 [Online]. Available: http://drops.dagstuhl.de/opus/volltexte/2017/7151.

Sommer, S., Camek, A., Becker, K., Buckl, C., Zirkler, A., Fiege, L., Armbruster, M., Spiegelberg, G., and Knoll, A. (2013). RACE: A centralized platform computer based architecture for automotive applications. *2013 IEEE International Electric Vehicle Conference (IEVC)*, Santa Clara, CA, pp. 1–6.

Traub, M., Maier, A., and Barbehon, K. (2017). Future automotive architecture and the impact of IT trends. *IEEE Software*, 34(3), 27–32.

Vdovic, H., Babic, J., and Podobnik, V. (2019). Automotive software in connected and autonomous electric vehicles: A review. *IEEE Access*, 7, p. 1.

Wicaksana, A., Charif, A., Andriamisaina, C., and Ventroux, N. (2019). Hybrid prototyping methodology for rapid system validation in HW/SW co-design. *Proceedings of Design and Architectures for Signal and Image Processing*.

6

Post-Moore Datacenter Server Architecture

Babak FALSAFI

School of Computer and Communication Sciences, EPFL, Lausanne, Switzerland

Cloud providers are building infrastructure at unprecedented speeds. We have witnessed the emergence of data-based information technology (IT) in almost every aspect of our life, from commerce, healthcare, entertainment and governance to scientific discovery. The demand for processing, communicating and storing data has grown faster than conventional growth in digital platforms. Meanwhile, the conventional silicon technologies that we have relied on for the past several decades, which led to the exponential growth in IT, have slowed down (the conventional 40% per year increase in density has dropped to 15% per year in recent years). While the increase in demand for data-centric IT continues and platform scalability has reached diminishing returns, the basic architecture of a modern server blade still dates back to the CPU-based desktop PC of the 1980s, which manages memory at hardware speeds but accesses the network and storage through the OS and legacy software stacks and peripheral interfaces. This chapter presents the promising recent approaches to redesigning the datacenter server blade's software and hardware of the future, in order to integrate heterogeneity into silicon and enable platform scalability with emerging network fabrics.

For a color version of all figures in this book, see www.iste.co.uk/andrade/multi2.zip.

Multi-Processor System-on-Chip 2 – Applications,
coordinated by Liliana ANDRADE and Frédéric ROUSSEAU.

6.1. Introduction

Today's IT services are provided by centralized infrastructure called datacenters. In contrast to supercomputers that are aimed at the high-cost/high-performance scientific domain, datacenters consist of low-cost servers for high-volume data processing, communication and storage. Datacenter owners prioritize capital and operating costs over ultimate performance (e.g. optimizing for higher performance per watt). Datacenters can vary in size and scale, with on-premise facilities consuming around 1 MW of electricity and large-scale warehouse-scale facilities consuming up to 250 MW of electricity, and costing several billions of dollars. The latter are at physical limits defined by sources of electricity, networking and cooling.

The emergence of datacenters is a direct response to a major IT paradigm shift to cloud computing and services, at the center of which is data. As individuals, we need ubiquitous access, exchange and sharing of data with those we interact with through cloud services. Similarly, businesses, governments and societies rely on cloud services to collect and analyze data, to exchange information to improve their products and services and, ultimately, to enhance our lives. Data now also lies at the core of the supply chain for both products and services in modern economies. Through analytics on massive datasets, data-centric science now complements theoretical, empirical and simulation-driven science as a fourth paradigm for scientific discovery. The net result is that our daily demands for storing, communicating and analyzing data are growing at much higher rates than improvements in semiconductor technology can sustain.

Recent years have also witnessed the emergence of a continuum of collaborating datacenters that vary in scale: the "edge" cloud provides services to users with proximity, real-time or sensitivity and privacy constraints, hosted on-premises; the "Internet exchange point (IXP)" cloud provides Internet services to enable connectivity; and "public" clouds hosted on warehouse-scale infrastructure offer globally available services, but potentially transcend the boundaries of data protection.

This unprecedented growth in the demand for datacenters has a direct impact on the slowdown in silicon scaling, the very phenomenon that has led to the proliferation of IT over the past five decades. Since the inception of silicon-based integrated circuits, digital platforms have enjoyed a doubling in density every two years (Moore's law) and in performance every 18 months with a roughly constant power envelope. Vendors have managed to limit the growth of the power envelope through lower transistor supply (and commensurately threshold) voltages, which is termed Dennard's law, while increasing the frequency and die size. The last two decades have witnessed a drastic slowdown in voltage scaling due to quantum effects, as the transistor channel has reached dimensions of the order of tens of nanometers, resulting in a dramatic shift towards platform designs that are more energy-efficient. Similarly, density scaling has slowed down, with a conventional scaling of 41% per year having now reached 17% per year (Brooks 2019) and must soon be coming to a halt. If the slowdown in silicon scaling is not mitigated, it will severely limit our ability to tap into the nearly limitless opportunities that massive data can offer.

While silicon scaling has slowed down, there are a number of technologies on the horizon that can help improve both the performance and capacity of datacenters. These technologies are not silver bullets that are likely to lead to sustained exponential improvements over half a dozen decades (as in Moore's law), but will nevertheless help improve cost and/or performance. Examples are die-stacking, from readily available interposer technologies to 3D monolithic chips, which will not only help with integration and density but also improve efficiency by limiting the distance the data has to move over a wire. Emerging memory technologies (e.g. PCM, STT-RAM and memristor) enable capacity scaling beyond DRAM. In networking, optical fabrics are projected to grow in bandwidth by 20% per year, which is a higher rate than projected improvements in silicon density, pushing platforms towards accommodating higher communication to computation ratios in software. Similarly, there are a number of emerging applications, including AI, massive data management and serverless computing, with highly diverse requirements for resource provisioning between processing, storage and communication, making it difficult to accommodate them through general-purpose computing platforms without silicon scaling.

In light of these conflicting trends in the demand for growth and conventional platform scaling, a complete rethinking of server architecture is needed to overcome density and efficiency challenges in datacenters with implications on all layers of the computation stack, up to algorithms. We envision platform scalability in the post-Moore era along three axes: (1) **I**ntegration, technologies for tighter integration of components (from algorithms to infrastructure) to minimize overhead in moving data, (2) **S**pecialization, technologies for accelerating critical services, mapping algorithms into representations that are closer to the platform and avoiding overhead in general-purpose processing and abstractions, and (3) **A**pproximation, technologies for gauging and tuning computational output quality to optimize resources used in computation. We refer to these axes as the ISA triangle of post-Moore server design.

6.2. Background: today's blades are from the desktops of the 1980s

The slowdown in silicon scaling and the continuing demands on IT growth require a full overhaul of the server architecture. Modern IT services are primarily online, hosting data that are directly in-memory to help deliver under tight latency constraints. Because these services are distributed over the network and network fabrics have improved in speed over the years, the key performance bottleneck in today's services is access to memory. Memory has conventionally been a passive component in the system, with access at hardware speeds only available to the CPU. Beyond CPU, any components (e.g. accelerators, networks, storage) wishing to access memory must request access through the host OS, which would then coordinate protected access among multiple software entities.

Today's blade servers inherit this CPU-based structure from the desktops of the 1980s. Figure 6.1 shows the organization of such a desktop. In these platforms, a CPU coordinates access to memory through a chipset at hardware speeds (e.g. a 33 MHz CPU accesses DRAM with a latency of 250 ns). The host OS, in turn, moves data between the storage and the network through an I/O bus. Because the desktop was primarily used as a multiprogrammed platform to run software from multiple users, the hardware/software mechanisms provided focused on isolating memory among users/software with infrequent activity in storage and disks. Figure 6.2 shows the organization of today's blade server, with one or two CPU sockets coordinating access to memory through integrated memory and cache coherence controllers at hardware speeds (e.g. 2 GHz CPUs access DRAM with a latency of 50 ns). The host OS moves data using legacy abstractions (e.g. address translation and paging) through a legacy interface (e.g. PCIe) from storage, network and discrete accelerators (e.g. GPUs and FPGAs).

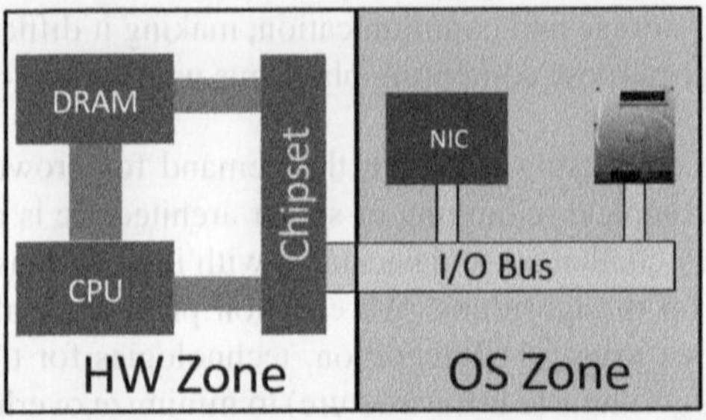

Figure 6.1. *The anatomy of a desktop in the 1980s*

This CPU-centric view of the world and the CPU domination of memory have led to a number of emerging trends. Because modern IT services are often memory-resident, all non-CPU components in the systems must go through the host CPU and OS to establish protected access to memory. Enabling direct access to memory still requires multiple levels of buffering between the OS, the middleware and the application, creating a performance bottleneck. Even bypassing the buffering would still require traversing the I/O interface, which remains a hardware bottleneck. In datacenters, the latter is of critical importance because improvements in network fabrics are moving in-memory services towards half a dozen tiers with μs-level handler functionality, exposing the bottleneck to memory. The net result is that components connected to the I/O interface integrate increasing levels of both logic and memory to decouple themselves from the host and implement services directly. The latter creates a fragmentation of silicon in a server node of not only logic but also memory, which is the most precious silicon resource for in-memory services.

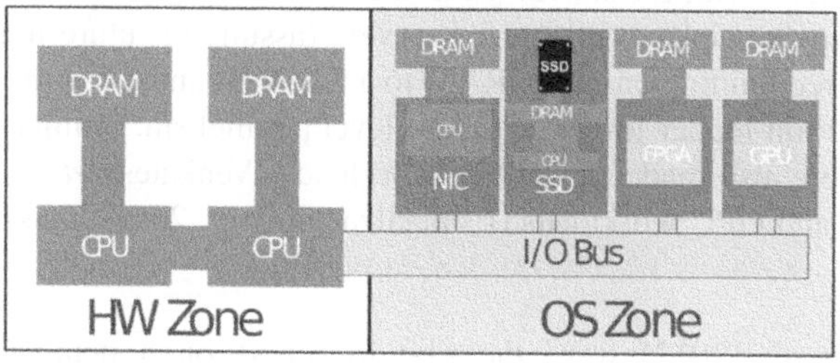

Figure 6.2. *The anatomy of a modern dual-socket server blade*

6.3. Memory-centric server design

While multi-processors laid the foundation for scientific computing and supercomputers, which almost date back to the beginning of the computing era, they only became commercially successful in the 1990s, in the form of shared-memory servers that run database workloads (e.g. OLTP or DSS). These workloads mostly benefitted from a high-performance memory system and were quite distinct in CPU resource utilization from both desktops (the dominant market for microprocessors) and scientific workloads (used in supercomputers). A myriad of papers in the literature in the 1990s evaluated the performance of server and database workloads on modern microprocessors, all of which concluded that there was a mismatch between the workload requirements and the microarchitectural mechanisms built in CPUs (Keeton *et al.* 1998; Ranganathan *et al.* 1998; Ailamaki *et al.* 1999). These studies revealed that server workloads, unlike desktop and HPC workloads, which mostly accessed memory through linked data structures, showed limited memory-level or arithmetic-level parallelism (with little use for floating-point arithmetic), but often had large instruction working sets due to the increased depth of the software stack.

With the slowdown in Dennard scaling and the emergence of multi-core servers in the 2000s, Hardavellas published a series of papers arguing that memory hierarchy features of commercial server-class multi-core CPUs diverged from the needs of commercial server workloads (including both data-based workloads and web workloads). Because of the large disparity between on-chip SRAM provisioning in caches (e.g. MBs) and off-chip DRAM provisioning for memory-resident workloads (e.g. GBs), server software stacks mostly benefitted from caching instructions on-chip with a capacity that was in the same order of magnitude (e.g. MBs) as the capacity of the on-chip cache hierarchy. In contrast to instructions, the vast disparity between on-chip cache and off-chip DRAM capacity limited any data reuse in caches for these workloads. Surprisingly, Hardavellas first showed that chips with larger last-level caches (LLCs) slowed down server workloads (Hardavellas *et al.* 2007) relative to those with smaller caches but with a faster instruction supply from those caches. He then showed, through a careful projection of technology, that servers were on the path to *dark silicon* with the slowdown in Dennard scaling (Hardavellas *et al.* 2011) and optimal provisioning of silicon for servers (assuming future higher bandwidth memory) required limiting on-chip SRAM to a few MBs and using the rest of the chip for cores to exploit higher levels of thread-level parallelism. Similar projections for dark silicon were also made for mobile workloads (Venkatesh *et al.* 2010), as well as for desktop and GPU workloads (Esmaeilzadeh *et al.* 2011), making the case for specialization.

Ferdman *et al.* (2012) were the first to report on a detailed study of the microarchitectural requirements of open-source server workloads in datacenters with x86 blade servers, who concluded that a key bottleneck was instruction supply, related to the overprovisioning of the LLC, which made it slower. With the recommendations for a correctly sized LLC to capture the instruction working set and lower out-of-order core complexity to match the degree of instruction-level parallelism in the threads to reduce the overall core area and power, they argued for many-core server processors that had an order of magnitude that was more silicon provisioned for cores than for LLCs, with a smaller, faster LLC and a fast supply network to the cores. Follow-on work at Google (Kanev *et al.* 2015; Ayers *et al.* 2019) corroborated the instruction supply bottleneck. Recent work also identified a key source of the bottleneck in the under-provisioning of BTB hierarchies in modern cores (Kumar *et al.* 2017), which have been mitigated in recent x86 server CPUs that allow fetch-directed prefetching of instructions from the LLC to hide the instruction supply latency.

Lotfi-Kamran *et al.* (2012a, b) followed these recommendations and derived optimal silicon provisioning for open-source server workloads based on the core area, LLC capacity and the on-chip network to reduce the distance from the cores to the LLC and accelerate on-chip instruction supply and data-miss detection. Figure 6.3 compares the anatomy of a conventional server CPU with a specialized server CPU for the same area of silicon. A custom many-core CPU with lower-complexity out-of-order cores and only L1 instruction and data caches without private L2 caches would enable a higher number of cores and degree of thread-level parallelism for the same area as conventional cores. Reclaiming an unused cache area (e.g. private L2s and a smaller LLC) can help dedicate silicon to more cores in order to exploit more thread-level parallelism. Because, in server workloads, the on-chip traffic is primarily core-to-LLC traffic with core requests supplied directly by the LLC (with instructions hitting and data often missing), core-to-core coherence requests account for less than 2% of the overall traffic. While crossbars are effective at low-latency, high-throughput connectivity among cores and the LLC, custom network-on-chip designs that are optimized for core-to-LLC communication rather than for core-to-core communication (Lotfi-Kamran *et al.* 2012a, b; Volos *et al.* 2012) further improve the overall latency, throughput, area and power over conventional networks. The first server-grade, 64-bit, ARM-based CPU from Cavium, ThunderX, with a higher ratio of core-to-LLC area provisioning, larger L1 instruction caches and a crossbar to improve instruction supply speed, was inspired by this work.

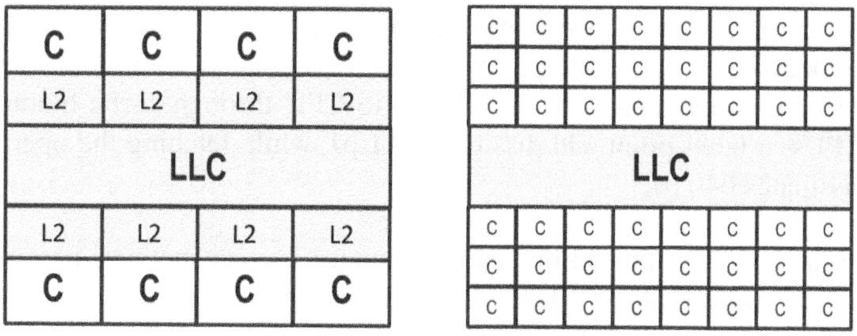

Figure 6.3. *Chip organization in an x86-based CPU (left) and a custom many-core CPU (right)*

6.4. Data management accelerators

The slowdown in Dennard scaling and Moore's law has also led to a myriad of accelerators. GPUs are the most common form of accelerators that have been retrofitted from graphics applications to large-scale programmable data-parallel applications, including deep learning. Other forms of accelerators, such as FPGAs, are now deployed in datacenters (e.g. Microsoft Azure), in order to implement a distributed machine learning service, and also to help with network and storage services, including encryption and compression. These accelerators are discrete, equipped with their own DRAM and connected to the CPU through the I/O interface.

Many datacenter services in scale-out datacenters perform simple data analytic operations (e.g. scan, lookup, join) in linked data structures in memory. While custom many-core CPUs improve silicon efficiency for such operations, programmable out-of-order cores are overkill for workloads that perform little arithmetic. Recent proposals advocate for programmable accelerators to traverse data structures and perform simple analytic operations. Walkers (Kocberber *et al.* 2013) was a hardware/software co-design to identify the minimal software with minimal mapping onto silicon in order to perform a database join near a core. Walkers used the existing address translation hardware from the core and saturated the available bandwidth to the L1 cache with the core idle while the accelerator executes. These results indicate a 15x throughput/watt improvement over an out-of-order core. Oracle also reported similar improvements in throughput/watt from RAPID (Balkesen *et al.* 2018) accelerating SQL analytics. An example of other near-CPU accelerators is the recent proposals to accelerate worst-case optimal join (WCOJ) for graph matching workloads (Kalinsky *et al.* 2020).

Near-memory processing has also emerged as a promising approach for data analytics. Recent advances in serial memory interfaces and interposer packaging have enabled the integration of logic near DRAM. Near-memory processing is less intrusive than in-memory processing, which was pioneered in the 1990s but was not adopted because of the extreme cost sensitivity of DRAM and the negative impact on density when integrating logic in DRAM. Simple analytic operations highly benefit from near-memory processing because the per-byte energy associated with these operations is dominated by data movement (from DRAM to CPU through cache hierarchies). In modern CPUs, a fixed-point add dissipates 0.1 pJ, while fetching the operands from DRAM dissipates 640 pJ.

We discussed the opportunities and challenges in near-memory processing in a recent opinion piece (Falsafi 2016). A key opportunity in near-memory processing is the use of streaming algorithms for analytics. Conventional CPU-based algorithms are designed for locality in reference and random memory access in cache hierarchies. Unlike caches, DRAM (and emerging storage-class memory) is internally optimized for density and behaves like a block device (e.g. a disk). Upon each access, an entire row of 1–2 KB of data is loaded into an internal buffer. Because loading the row accounts for much of the latency and energy for each access, streaming the row out of DRAM would lead to the optimal use of DRAM in terms of efficiency. Surprisingly, streaming algorithms for analytics that have a higher order of complexity (i.e. to compute more), such as sort join, not only execute faster but also dissipate less energy near-memory when compared to CPU-based algorithms such as hash join. The Mondrian Data Engine (Drumond *et al.* 2017) exhibits an algorithm/microarchitecture co-design for Spark in-memory analytic operators. Figure 6.4 shows a log scale of speedup for a simple scan, sort, groupby and join, the underlying operators in Spark analytics. The graph shows that streaming near-memory alone with a many-core CPU (NMP-OoO) achieves a 10x speedup over a conventional CPU and memory with cache hierarchies. A custom microarchitecture for streaming with wide SIMD units increases the speedup to 50x.

6.5. Integrated network controllers

The conventional placement of network (and storage) controllers across the legacy I/O interfaces (Figure 6.2) is also inherited from the desktop in the 1980s, which not only poses a fundamental impediment to efficiency, but also fragments memory (the most precious resource) in servers and datacenters. Recent intelligent network controllers include multi-core out-of-order ARM cores that reach GBs of DRAM, isolated from the host by the I/O bus, placing the CPU host and OS on the critical path of access to host memory.

However, many have advocated the integration of the network controller logic into the CPU socket (Binkert *et al.* 2006). Scale-Out NUMA (Novakovic *et al.* 2014) has been proposed for integrating minimal network logic into the coherence domain of a CPU socket, in order to implement RDMA-like verbs for user-level reads, writes and atomics (Daglis *et al.* 2016) with protected access to remote memory. The latter enables the pooling of memory across a rack of servers to help with load balancing data services with skewed request distributions and analytics (e.g. graph processing) with an inherent skew in data partitions. A novel integrated network controller microarchitecture has also been shown to scale with the number of CPU cores (Daglis *et al.* 2015).

With network fabrics slated to improve in bandwidth by 20% per year and the push towards microservices, the network communication software stack becomes a bottleneck. Modern microservices communicate through the RPC protocol to run computation on data that is hosted on a remote node. An integrated network controller can also help with load balancing RPC requests, monitoring the RPC handler load directly on the socket and dispatching handlers with cache affinity information about handler metadata (Daglis *et al.* 2019). Microservices are also written in diverse software environments with varying protocols in order to structure objects in memory. Recent reports from the industry have indicated that data transformation upon communication between datacenter services accounts for a larger fraction of overall CPU cycles (e.g. Google's ProtoBuf accounts for 10% of all cycles). Hardware/software co-design for data transformation with integrated network controllers can help to eliminate the overhead and perform transformation at the network line rate (Pourhabibi *et al.* 2020).

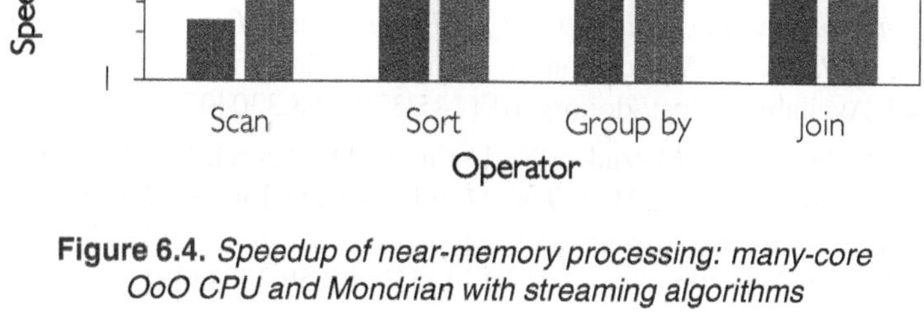

Figure 6.4. *Speedup of near-memory processing: many-core OoO CPU and Mondrian with streaming algorithms*

6.6. References

Ailamaki, A., DeWitt, D.J., Hill, M.D., and Wood, D.A. (1999). DBMSs on a modern processor: Where does time go? *Proceedings of the 25th International Conference on Very Large Data Bases, VLDB'99*. Morgan Kaufmann Publishers Inc., San Francisco, pp. 266–277.

Ayers, G., Nagendra, N.P., August, D.I., Cho, H.K., Kanev, S., Kozyrakis, C., Krishnamurthy, T., Litz, H., Moseley, T., and Ranganathan, P. (2019). AsmDB: Understanding and mitigating front-end stalls in warehouse-scale computers. *Proceedings of the 46th International Symposium on Computer Architecture, ISCA'19*, Association for Computing Machinery, New York, pp. 462–473 [Online]. Available: https://doi.org/10.1145/3307650.3322234.

Balkesen, C., Kunal, N., Giannikis, G., Fender, P., Sundara, S., Schmidt, F., Wen, J., Agrawal, S., Raghavan, A., and Varadarajan, V., Viswanathan, A., Chandrasekaran, B., Idicula, S., Agarwal, N., and Sedlar, E. (2018). RAPID: In-memory analytical query processing engine with extreme performance per watt. *Proceedings of the 2018 International Conference on Management of Data, SIGMOD'18*, Association for Computing Machinery, New York, pp. 1407–1419 [Online]. Available: https://doi.org/10.1145/3183713.3190655.

Binkert, N.L., Saidi, A.G., and Reinhardt, S.K. (2006). Integrated network interfaces for high-bandwidth TCP/IP. *SIGOPS Operating Systems Review*, 40(5), 315–324 [Online]. Available: https://doi.org/10.1145/1168917.1168897.

Brooks, D. (2019). What's the future of technology scaling? [Online]. Available: https://www.sigarch.org/whats-the-future-of-technology-scaling/.

Daglis, A., Novakovic, S., Bugnion, E., Falsafi, B., and Grot, B. (2015). Manycore network interfaces for in-memory rack-scale computing. *Proceedings of the 42nd International Symposium on Computer Architecture, ISCA'15*, pp. 567–579 [Online]. Available: https://doi.org/ 10.1145/2872887.2750415.

Daglis, A., Ustiugov, D., Novakovic, S., Bugnion, E., Falsafi, B., and Grot, B. (2016). SABRes: Atomic object reads for in-memory rack-scale computing. *Proceedings of the 49th Annual IEEE/ACM International Symposium on Microarchitecture*.

Daglis, A., Sutherland, M., and Falsafi, B. (2019). RPCValet: NI-driven tail-aware balancing of µs-Scale RPCs. *Proceedings of the Twenty-Fourth International Conference on Architectural Support for Programming Languages and Operating Systems, ASPLOS'19*, Association for Computing Machinery, New York, pp. 35–48 [Online]. Available: https://doi.org/10.1145/3297858.3304070.

Drumond, M., Daglis, A., Mirzadeh, N., Ustiugov, D., Picorel, J., Falsafi, B., Grot, B., and Pnevmatikatos, D. (2017). The Mondrian Data Engine. *Proceedings of the 42nd International Symposium on Computer Architecture, ISCA'17*, pp. 639–651 [Online]. Available: https://doi.org/ 10.1145/3140659.3080233.

Esmaeilzadeh, H., Blem, E., St. Amant, R., Sankaralingam, K., and Burger, D. (2011). Dark silicon and the end of multicore scaling. *Proceedings of the 38th Annual International Symposium on Computer Architecture, ISCA'11*, Association for Computing Machinery, New York, pp. 365–376 [Online]. Available: https://doi.org/10.1145/2000064.2000108.

Falsafi, B. (2016). Near-memory data services. *IEEE Micro*, 36(1), 6–13 [Online]. Available: https://doi.org/10.1109/MM.2016.9.

Ferdman, M., Adileh, A., Kocberber, O., Volos, S., Alisafaee, M., Jevdjic, D., Kaynak, C., Popescu, A.D., Ailamaki, A., and Falsafi, B. (2012). Clearing the clouds: A study of emerging scale-out workloads on modern hardware. *SIGPLAN Notices*, 47(4), 37–48 [Online]. Available: https://doi.org/10.1145/2248487.2150982.

Hardavellas, N., Pandis, I., Johnson, R., Mancheril, N., Ailamaki, A., and Falsafi, B. (2007). Database servers on chip multiprocessors: Limitations and opportunities. *Proceedings of the Third Biennial Conference on Innovative Data Systems Research, CIDR'07*, pp. 79–85.

Hardavellas, N., Ferdman, M., Falsafi, B., and Ailamaki, A. (2011). Toward dark silicon in servers. *IEEE Micro*, 31(4), 6–15 [Online]. Available: https://doi.org/10.1109/MM.2011.77.

Kalinsky, O., Kimelfeld, B., and Etsion, Y. (2020). The TrieJax architecture: Accelerating graph operations through relational joins. *Proceedings of the Twenty-Fifth International Conference on Architectural Support for Programming Languages and Operating Systems, ASPLOS'20*, Association for Computing Machinery, New York, pp. 1217–1231.

Kanev, S., Darago, J.P., Hazelwood, K., Ranganathan, P., Moseley, T., Wei, G.-Y., and Brooks, D. (2015). Profiling a warehouse-scale computer. *Proceedings of the 42nd Annual International Symposium on Computer Architecture, ISCA'15*, Association for Computing Machinery, New York, pp. 158–169 [Online]. Available: https://doi.org/10.1145/2749469.2750392.

Keeton, K., Patterson, D.A., He, Y.Q., Raphael, R.C., and Baker, W.E. (1998). Performance characterization of a Quad Pentium Pro SMP using OLTP workloads. *Proceedings of the 25th International Symposium on Computer Architecture, ISCA'98*, pp. 5–26 [Online]. Available: https://doi.org/10.1145/279361.279364.

Kocberber, O., Grot, B., Picorel, J., Falsafi, B., Lim, K., and Ranganathan, P. (2013). Meet the walkers: Accelerating index traversals for in-memory databases. *Proceedings of the 46th Annual IEEE/ACM International Symposium on Microarchitecture, MICRO-46*, Association for Computing Machinery, New York, pp. 468–479 [Online]. Available: https://doi.org/10.1145/2540708.2540748.

Kumar, R., Huang, C., Grot, B., and Nagarajan, V. (2017). Boomerang: A metadata- free architecture for control flow delivery. *Proceedings of the IEEE International Symposium on High Performance Computer Architecture, HPCA'17*, IEEE Computer Society, Austin, pp. 493–504 [Online]. Available: https://doi.org/10.1109/HPCA.2017.53.

Lotfi-Kamran, P., Grot, B., and Falsafi, B. (2012a). NOC-Out: Microarchitecting a scale-out processor. *Proceedings of the 2012 45th Annual IEEE/ACM International Symposium on Microarchitecture, MICRO-45*, IEEE Computer Society, pp. 177–187 [Online]. Available: https://doi.org/10.1109/MICRO.2012.25.

Lotfi-Kamran, P., Grot, B., Ferdman, M., Volos, S., Kocberber, O., Picorel, J., Adileh, A., Jevdjic, D., Idgunji, S., Ozer, E., and Falsafi, B. (2012b). Scale-out processors. *Proceedings of the 39th International Symposium on Computer Architecture, ISCA'12*, pp. 500–511 [Online]. Available: https://doi.org/10.1145/2366231.2337217.

Novakovic, S., Daglis, A., Bugnion, E., Falsafi, B., and Grot, B. (2014). Scale-out NUMA. *Proceedings of the Nineteenth International Conference on Architectural Support for Programming Languages and Operating Systems, ASPLOS'14*, Association for Computing Machinery, New York, pp. 3–18 [Online]. Available: https://doi.org/10.1145/2541940.2541965.

Pourhabibi, A., Gupta, S., Kassir, H., Sutherland, M., Tian, Z., Paulo Drumond, M., Falsafi, B., and Koch, C. (2020). Optimus prime: Accelerating data transformation in servers. *Proceedings of the Twenty-Fifth International Conference on Architectural Support for Programming Languages and Operating Systems, ASPLOS'20*, Association for Computing Machinery, New York, pp. 1203–1216.

Ranganathan, P., Gharachorloo, K., Adve, S.V., and Barroso, L.A. (1998). Performance of database workloads on shared-memory systems with out-of-order processors. *Proceedings of the Eighth Conference on Architectural Support for Programming Languages and Operating Systems, ASPLOS VIII*, Association for Computing Machinery, New York, pp. 307–318.

Venkatesh, G., Sampson, J., Goulding, N., Garcia, S., Bryksin, V., Lugo-Martinez, J., Swanson, S., and Taylor, M.B. (2010). Conservation cores: Reducing the energy of mature computations. *Proceedings of the Fifteenth International Conference on Architectural Support for Programming Languages and Operating Systems, ASPLOS XV*, Association for Computing Machinery, New York, pp. 205–218 [Online]. Available: https://doi.org/10.1145/1736020.1736044.

Volos, S., Seiculescu, C., Grot, B., Khosro Pour, N., Falsafi, B., and De Micheli, G. (2012). CCNoC: Specializing on-chip interconnects for energy efficiency in cache-coherent servers. *2012 Sixth IEEE/ACM International Symposium on Networks-on-Chip (NoCs)*, IEEE Computer Society, Copenhagen, pp. 67–74 [Online]. Available: https://doi.org/10.1109/NOCS.2012.15.

PART 3

Architecture Examples and Tools for MPSoC

7

SESAM: A Comprehensive Framework for Cyber–Physical System Prototyping

Amir CHARIF[1], Arief WICAKSANA[2], Salah-Eddine SAIDI[1], Tanguy SASSOLAS[1], Caaliph ANDRIAMISAINA[1] and Nicolas VENTROUX[1]

[1] *LIST, CEA, Gif-sur-Yvette, France*

[2] *TIMA Lab, Université Grenoble Alpes, France*

Because systems may consist of several multi-core SoCs (Systems-on-Chip), it is no longer relevant to consider a single SoC as the full system, during design and evaluation phases. Today, CPS (Cyber–Physical Systems) not only incorporate several full computers, but are also very tightly coupled with the physical environment, for example autonomous cars. This chapter is an introduction to SESAM, a comprehensive platform for the prototyping of modern CPS, and systems of systems, in general. SESAM accelerates and facilitates the prototyping of CPS by adopting a hybrid prototyping approach, composed of subsystems from various domains and levels of abstraction. For instance, virtual subsystems from QEMU and SystemC, RTL designs, as well as physical and analog FMI models, can be integrated and connected together at a high level, to quickly build fast prototypes from their specification. SESAM also ships with a set of verification tools that help us to assess critical aspects of the system, including reliability and power consumption.

For a color version of all figures in this book, see www.iste.co.uk/andrade/multi2.zip.

Multi-Processor System-on-Chip 2 – Applications,
coordinated by Liliana ANDRADE and Frédéric ROUSSEAU.

7.1. Introduction

CPS are becoming key to many fast growing markets, from consumer electronics with IoT devices interacting strongly together and with their environment, to safety critical industries such as the upcoming autonomous car with its variety of sensors and controls. The emergence of CPS brings the challenges of SoC designs to yet another complexity level. Indeed, CPS are electronic systems with hardware and software that interact with their environment through sensors and actuators, thus requiring the ability to encompass their design in a holistic manner. The cost and effort involved in the design and implementation of such systems can quickly escalate, compelling system designers and software developers to rely on early prototypes for the thorough evaluation of their designs, in terms of correctness, performance, scalability, reliability, etc.

Of course, the prototyping of such complex systems requires tools that are powerful enough to take on a number of challenges. First, the tool must be designed to build and compose several complex subsystems from the ground up; otherwise, users may find themselves wasting precious time in devising *ad hoc* interfaces to stitch multiple prototypes together. Second, the prototyping tool should scale properly, in terms of speed, to allow rapid execution and iteration over candidate designs. As no single existing tool can cover all the necessary building blocks for all future systems, relevant tools should not only be extensible, but also be interoperable with other tools through common standard interfaces. Finally, prototyping CPS requires seamless interaction between the digital and analog worlds.

This chapter introduces SESAM, a comprehensive framework designed to enable and accelerate electronic and cyber–physical system prototyping throughout its design phases. The chapter starts with an overview of the various building blocks of the SESAM framework in section 7.2, with a specific focus on extra-functional property assessment. The major part of this chapter then dwells on functional prototyping solutions provided by SESAM. First, digital system prototyping solutions are presented, starting with virtual prototyping in section 7.3, and then FPGA-virtual hybrid prototyping in section 7.4. Finally, our solution for cyber–physical co-simulation is described in section 7.5 before concluding in section 7.6.

7.2. An overview of the SESAM platform

SESAM is an integrated EDA framework for complex electronic systems ranging from cyber–physical systems to microservers. SESAM provides a holistic environment to address HW/SW co-design, exploration and validation through virtual prototyping, HW prototyping and emulation while taking into account power, temperature and reliability factors. The virtual prototyping tool, named VPSim

(Charif *et al.* 2019), was specifically designed to accelerate software/hardware co-validation at the earliest design stages of all kinds of computer architectures. It can be used to easily compose, simulate and explore new hardware architectures, to run, profile and debug full software stacks on the simulated platforms, as well as, for example, to co-simulate with multi-physics tools through FMI. SESAM also comes with SCale (Ventroux and Sassolas 2016; Busnot *et al.* 2020), a parallel SystemC kernel, to accelerate any virtual prototyping solutions based on SystemC (including VPSim). It allows parallel, deterministic and standard compliant SystemC simulations of any TLM models, including those using the loosely timed coding style together with the DMI interface. It relies on a conflict-avoidance heuristic that avoids the vast majority of conflicts, combined with a conflict-detection and a fast rollback mechanism.

In addition, SESAM offers an FPGA-instrumented framework for non-intrusive internal debugging of FPGA prototypes. Hybrid simulation composed of VPSim and FPGA boards is then possible, to validate RTL designs inside a virtual prototyping environment. Finally, SESAM offers different solutions for the power and aging estimation and modeling, as well as fault injection at different levels of abstractions. There is no existing equivalent framework in the state-of-the-art. Only commercial EDA vendors such as Synopsys or Mentor Graphics provide a complete set of tools for silicon design and verification. For instance, by combining Virtualizer, HAPS, Z01X and SpyGlass (Synopsys 2019), all the features brought by SESAM could be covered.

7.2.1. *Multi-abstraction system prototyping*

These days, two independent methods based on different abstraction levels are commonly used for SoC prototyping: transaction-level model (TLM)-based virtual prototyping and FPGA-based prototyping. Virtual prototyping permits the acceleration of pre-RTL software development, by leveraging the speed of TLM execution and providing more efficient debug and analysis scenarios. On the other hand, FPGA-based prototyping brings cycle accuracy together with high-performance execution later in the design flow. SESAM provides the possibility of coupling these two methods (Wicaksana *et al.* 2019) throughout the design cycles.

With the increase of SoC complexity, it is often necessary to perform multi-FPGA-based prototyping. Before the multi-FPGA prototyping, a design partitioning has to be performed. To accelerate the partitioning, which is a complex and time-consuming task, SESAM also provides an automatic partitioning tool. Currently, the tool takes as input a design netlist and performs partitioning, by taking into account, several constraints related to the platform such as FPGA resource limitations, inter-FPGA interface limitations, etc.

7.2.2. *Assessing extra-functional system properties*

7.2.2.1. *Power modeling*

Taking into account the power consumption of a SoC at the earliest design stages is key for addressing the stringent power budget expected by the CPS market. This requires us to be able to evaluate its behavior throughout the evolution of the architecture as well as of its software and power management strategy. Thus, an efficient power modeling must be sought, to obtain accurate power evaluation during prototyping, while not hampering the functional prototyping performance.

With SESAM, it is possible to automatically generate a power model through the use of a learning-based methodology. The proposed power modeling methodology is based on the hardware performance monitors (HPM). These HPM monitor different events that take place during instruction execution and reveal large amounts of information about power consumption. These events can be the number of cycles, the number of retired instructions, cache misses, etc. The power model generation flow, takes as input, the activity of hardware performance monitors and a dynamic power consumption reference, which can be obtained with a signoff power analysis tool such as Primetime PX, and performs a linear regression to generate the power model.

7.2.2.2. *Reliability*

Aging of integrated circuits is recognized as a major reliability issue, and the estimation of its effects is mandatory in the design of safety-critical systems. Fault injection is a valid method to evaluate the effects of faults on systems (Sandionigi and Heron 2019).

SESAM integrates a fault injection framework to evaluate the effects of oxide breakdown on the paths of integrated circuits. The fault injection methodology behind the framework is composed of three steps: fault injection campaign definition, fault injection execution and fault propagation analysis. The first step is based on a model that accurately describes the behavior of oxide breakdown. A two-level model is proposed: the oxide breakdown is first studied at the gate level for more accuracy, and is then translated to RTL (Register Transfer Level), to allow rapid fault injection. In the execution step, the fault injection is performed on the registers identified as more sensitive to oxide breakdown. The fault injection is done either by simulation or emulation at RTL. In the analysis step, the propagation of the faults injected into the system is observed to identify which faults become observable errors, which ones propagate only partially and are not observable (silent errors) and which ones do not propagate.

7.3. VPSim: fast and easy virtual prototyping

VPSim is a platform for building, assembling and driving the simulation of virtual prototypes. It was previously introduced in Charif *et al.* (2019). As summarized in

Figure 7.1, VPSim starts by taking a high-level description of the HW/SW platform, either in the Python language or from a user interface. VPSim uses this description to generate a SystemC simulation from the available models, which are either built into the default library of SystemC models, or provided by third-party libraries. During simulation, the user can inspect, debug and control the execution of software on the virtual platform. At the end of the execution, detailed statistics are provided to the user for architectural exploration, optimization, etc.

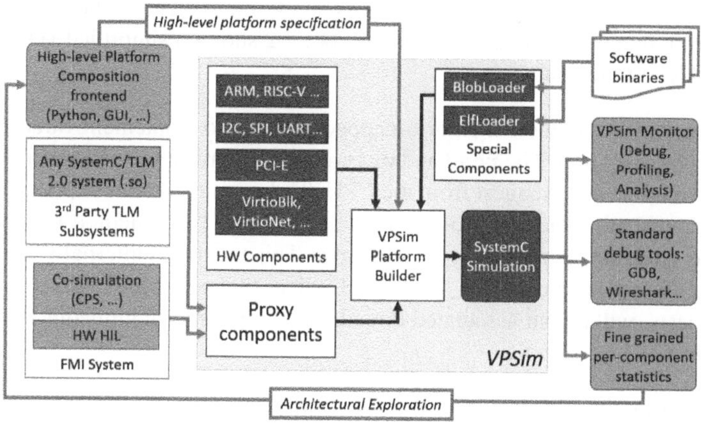

Figure 7.1. *Overview on the operation of VPSim*

In this section, we detail the latest key features brought to VPSim.

7.3.1. *Writing peripherals in Python*

In our previous work (Charif *et al.* 2019), we introduced the platform description framework of SESAM, which used the Python programming language. This methodology saves developers precious time when creating or reconfiguring virtual platforms from existing components (e.g. for design space exploration (DSE)).

We now push this idea even further, by allowing TLM slave and master components themselves, to be developed dynamically using the Python language. Of course, for performance considerations, this approach is aimed at developing peripheral devices and system registers rather than critical architectural components such as CPUs or memory. Indeed, as peripherals are seldom used within a simulation, their evaluation speed has a very limited impact on the overall simulation duration. The idea is to reduce the time spent in writing peripheral components and also increase productivity by giving developers access to all the powerful features and libraries of the Python language.

A Python device can be instantiated and connected to the rest of the system in the platform description, just like any other VPSim component. The following line instantiates a memory-mapped Python device and connects it to the system bus:

```
sysbus » PythonMMIO(classname='PL011Uart',
    base_address=0xff000000, size=0x1000, irq=0x22)
```

The `classname` attribute of the `PythonMMIO` component points to the name of the Python class that contains the implementation of the device. The other attributes are similar to the other memory-mapped devices in VPSim, indicating their address space and interrupt mappings. As an example, Figure 7.2 shows a simplified version of the `PL011` UART implementation.

VPSim expects Python memory-mapped devices to implement three methods: `read`, `write` and `loop`. The `read` and `write` methods are called whenever the device receives a read or write request from the CPU. The `read` method is expected to return a value, representing the result of the read access. The `loop` method is what allows the user to schedule operations in the simulator. This method is internally called in the context of a SystemC thread (`SC_THREAD`). Devices that only need to react to read and write requests, without an associated thread, can simply return immediately from the loop method.

To call back the simulator, the Python device can use a set of built-in functions, defined in the pyvp module (see Figure 7.2). All the functions must include a reference to the Python device as a first parameter. The most important functions are:

– `wait(device, nanoseconds)`, which yields control to the SystemC kernel for the specified amount of time;

– `interrupt(device, irqLine, irqValue)`, which raises or lowers an interrupt on the given line;

– `dma_read(device, address, size)` and `dma_write(device, address, size, value)`, which can be used to issue a memory access request from the device.

When modeling new platforms, this approach can be a real time saver, as system-specific registers that are required to run certain boot sequences can directly be implemented in Python along with the platform description. Only performance-critical components need to have a compiled SystemC implementation.

7.3.2. *The ModelProvider interface*

As a virtual prototyping solution, VPSim ships with a library of SystemC device models. However, for applications requiring a processor or peripherals that are not included in its library, VPSim must have the ability to import these models from external tools. To this end, VPSim defines a common interface of functions and callbacks named `ModelProvider`, to be implemented in external tools, to make them controllable from the VPSim infrastructure.

```
import pyvp
import sys
import tty
import select
import os

class PL011Uart:
    def __init__(self, **kwargs):
        self.base=kwargs["base"]
        self.irq=kwargs["irq"]
        # ... other initializations

        print("Initialized device pl011... (base=%s,irq=%s)" %
            (self.base,self.irq))

    def read(self, addr, size):
        off = addr - self.base
        self.update_status()
        if off == 0x00:
            return self.get_rx()
        elif off == 0x18:
            return self.get_status()
        elif off == 0x3C:
            return self.get_interrupt_unmasked()
        elif off == 0x40:
            return self.get_interrupt_masked()
        else:
            # other cases...

    def write(self, addr, value, size):
        off=addr-self.base
        if off == 0x00:
            self.backend_writechar(chr(value))
        elif off == 0x38:
            self.update_interrupt_enable(value)
        elif off == 0x44:
            self.upadte_interrupt_disable(value)
        else:
            # other cases...

        self.update_status()
        self.check_interrupt()

    def check_interrupt(self):
        if self.shouldInterrupt():
            pyvp.interrupt(self, self.irq, 1)
        else:
            pyvp.interrupt(self, self.irq, 0)

    def loop(self):
        while True:
            pyvp.wait(self, int(self.nsPerChar))
            self.update_status()
            self.check_interrupt()
```

Figure 7.2. *Simplistic implementation of the PL011 UART in Python*

When a tool implements the `ModelProvider` C interface, as done with QEMU (see section 7.3.3), the user can have full control over the tool directly from VPSim's Python interface. First, the external tool is loaded as a shared library into VPSim. Then, any processor or peripheral models available in the external tool can be instantiated as VPSim components of type `ModelProviderCPU` or `ModelProviderDev`, and then connected to any input/output port in the system. If the external tool takes options as command line parameters, these can be defined by the user through special components of type `ModelProviderParam1` or `ModelProviderParam2`. We will later illustrate their usage in the case of QEMU in section 7.3.3.1.

7.3.3. *QEMU support*

Although SESAM can support any external model provider (e.g. ARM Fast Models (Rodman 2008), Open Virtual Platforms (OVP), etc.), special attention is given to the open source system emulator QEMU, which supports a large variety of processor models for building ARM- or RISC-V-based virtual subsystems. QEMU (Bellard 2005) offers the fastest DBT (Dynamic Binary Translation)-based CPU emulation available, especially following the introduction of the multi-threaded tiny-code generator (MTTCG), which allows guest CPUs to leverage physical threads on the host machine. In addition, QEMU has a very large library of peripheral device models, which existing QEMU SystemC integrations typically fail to fully leverage.

Unfortunately, QEMU lacks the flexibility and accuracy required in HW/SW co-design. Building custom machines in QEMU is very difficult and requires a great deal of C development and debugging. Our integration makes it possible to describe complete subsystems in QEMU, using the default Python platform description language of VPSim. QEMU aims to perform functionally correct ultra-fast emulations and completely disregards architectural details, such as the memory hierarchy, which are key to certain applications. We make it possible for CPUs in QEMU to efficiently interact with a detailed memory hierarchy model written in SystemC. Finally, we bring essential fixes and improvements to some QEMU features, that are either not completely implemented or that have inaccurate implementations, as explained later in this subsection.

7.3.3.1. *Integration methodology*

In typical virtual prototyping solutions that make use of QEMU, two cases are common. In the first approach, only CPU models are wrapped and used in simulation. In the second approach, complete subsystems are manually built in QEMU, and the C code must be changed anytime the platform needs to be modified. We want to challenge these modes of operation.

By implementing the `ModelProvider` interface in QEMU, our goal is to allow everything in QEMU to be controlled from the same Python description used for the rest of the system. That is, building platforms using QEMU is no longer restricted to QEMU experts, but can be done by the users themselves.

Our current integration makes it possible to instantiate any internal device in QEMU, and connect it to any compatible SystemC port. For instance, to create a PL011 UART inside QEMU, the ModelProviderDev class, as described previously, can be used as follows in the Python description:

```
uart = ModelProviderDev(provider='qemu', model='pl011',
        base_address=0xff000000, irq=0x22)
```

The `uart` object can then be used as any other component in VPSim, for instance to connect it to the system bus:

```
bus >> uart
```

That is, complete new platforms can be created using QEMU devices, without writing a single C line of code. Any platform can be instantiated using one same binary, so no code generation, i.e. no compile time, is ever involved. Moreover, we make it possible to configure QEMU from the same Python description, using the well-known command line parameters. A command line parameter can also be created as a Python object of class `ModelProviderParam1` (for True/False options) or `ModelProviderParam2` (for options with a value). For example, to turn on debug mode:

```
ModelProviderParam2(provider='qemu', option='-gdb',
        value='tcp::10000')
```

Main memory accesses in QEMU are instrumented to be simulated by a detailed SystemC simulator of the memory hierarchy. This instrumentation can be disabled by the user to maximize the execution speed. This can be similarly performed using a qemu option passed through the `ModelProviderParam2`.

7.3.3.2. *Improvements to QEMU*

Several improvements had to be made to QEMU to provide a useful environment for co-design. First, one IP that is very useful is the ARM performance monitoring unit (PMU). The PMU gives user-space programs access to performance counters, which makes it possible to profile programs directly in the simulated Linux user-space using popular utilities such as perf. Moreover, some applications require access to several counters to optimize runtime. At the time of writing this, QEMU supports a few basic PMU counters, which we imported to VPSim and further completed with several useful counters from the memory hierarchy (cache miss and write-backs, etc.). The counters we implemented are exclusive to VPSim and cannot be upstreamed due to mainline QEMU not modeling the memory hierarchy.

Second, the way in which QEMU models time cannot be used as is for performance evaluation, even when using the icount (Instruction Count) mode of QEMU. We had to fully re-design time calculation for CPUs in QEMU, while also taking into account the impact of the memory hierarchy.

7.3.4. *Online simulation monitoring*

VPSim internally instantiates a special component called the VPSim Monitor. The Monitor allows the user to freeze the simulation during runtime, in order to collect statistics, inspect memory contents, schedule a new profiling session or even reconfigure components of the platform, without rebooting the system or ending the simulation. The Monitor comes in two variants. The first variant is non-intrusive to the simulation and is presented to the user in a separate console. The user can enter commands in this console to control the simulation. However, this variant has no notion of the simulated software.

A more flexible, but intrusive variant, is accessed from the simulated Linux user space as a regular piece of software called `sesam`. This program takes care of communicating with the `Monitor` component and therefore also has full control over the simulator; only this time, it also has access to the simulated software environment. The user-space variant program can, for instance, be used to collect statistics following the execution of a specific user application, making it possible to perform the entire exploration in one simulated user-space session directly, using scripts in the simulated session. All component configurations impacting performance but not functionality can be adapted at runtime (e.g. cache size or policy), but changes that impact the memory map or the number of components naturally require another execution.

7.3.5. *Acceleration methods*

To respond to current industrial needs in terms of performance, VPSim uses various techniques to accelerate simulations. VPSim adopts parallelism at various levels. First, it is possible to simulate CPU cores in parallel by using QEMU's MTTCG or the SCale parallel SystemC kernel (Ventroux and Sassolas 2016).

Second, VPSim effectively removes the memory hierarchy model from the critical path of the simulation, by simulating the architectural details concurrently with the functional execution. Unlike similar solutions (Kang *et al.* 2016; Deschamps *et al.* 2019), the delays computed by the memory hierarchy model are accounted for in the main simulation, resulting in negligible loss in accuracy compared to the sequential implementation. Moreover, our method is also able to model instruction caches at full accuracy, with very little impact on execution performance.

7.4. Hybrid prototyping

In addition to fast functional simulation and rapid software validation, VPSim also provides the possibility of system co-simulation/co-emulation in hardware/software (HW/SW) co-design (Wicaksana *et al.* 2019). We propose a hybrid prototyping methodology, which combines execution on VPSim with an RTL simulator, or an FPGA, to validate the system at various stages of development based on the availability of TLM and RTL implementations. As a result, SoC architects (users) are offered an optimized and flexible validation flow as well as the possibility of incremental architecture explorations.

The high complexity of modern SoC requires hardware and software parts to be developed and validated together in a condition close to the real application. The proposed hybrid prototyping methodology is able to address this challenge. For instance, developing a Linux driver for a new peripheral device requires access to a specific library and functions. The use of VPSim enables functional and compatibility verification of this driver directly on the targeted operating system. The co-simulation/co-emulation extension in VPSim allows us to reuse stable IPs or incrementally add new IPs to the architecture. As more SoCs integrate accelerators, the proposed hybrid prototyping framework can be an excellent tool in the development of such a system. The presented methodology can be used throughout the development of complex systems, from the early stage where the entire system is described only in functional models, until the final prototyping stage before manufacturing.

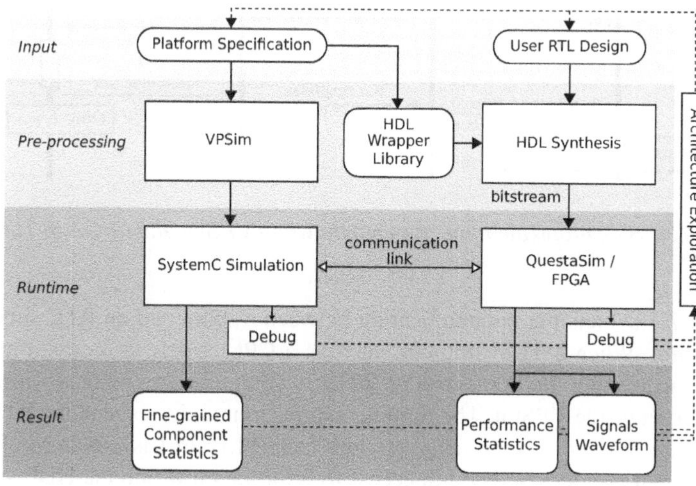

Figure 7.3. *System validation flow using the hybrid prototyping solution*

For such purposes, the proposed hybrid prototyping framework supports the system validation in HW/SW co-design. Figure 7.3 describes an overview of the validation flow offered by the framework. We propose to concurrently execute both virtual platforms in VPSim and RTL in the framework and to synchronize every quantum as per a loosely timed model in TLM. The framework takes as input, a platform specification, which describes the full system architecture and software stacks. It also takes the user's RTL design for the parts that need to be validated on an RTL simulator, such as QuestaSim, or on an FPGA. For the output, the framework generates the HW/SW validation results and performance statistics of each component.

7.4.1. *Co-simulation mode*

The co-simulation mode enables HW/SW validation after the RTL implementation is available in the development phase. This approach is supported by VPSim thanks to the implementation of TLM Remote Target in VPSim library. Remote Target is a proxy component, which is specially developed to connect VPSim with external prototyping devices, for example QuestaSim or FPGA. It forwards the read and write requests as well as interrupts between SystemC and RTL simulations. A TCP-based connection is used in the Remote Target to allow seamless communication and flexible RTL on the external side of VPSim.

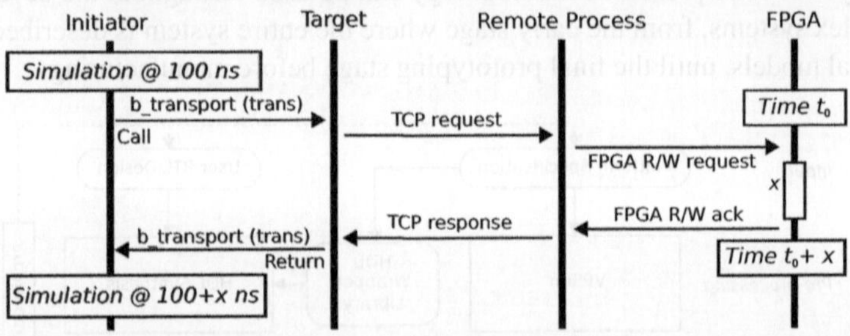

Figure 7.4. *Synchronization mechanism between VPSim and FPGA in TLM R/W*

Figure 7.4 depicts the communication between VPSim and an RTL simulator in co-simulation. When a TLM initiator, such as a CPU, needs to access a peripheral device or accelerator implemented in RTL, it communicates with an instantiated Remote Target inside VPSim. The Remote Target communicates with the RTL design in QuestaSim through foreign language interface (FLI). FLI is a feature offered by QuestaSim that provides procedural access to information within the HDL simulator, via C programming language. In this mode, the write and read process in the simulation is synchronously driven by the TLM initiator in VPSim. The environment of VPSim and QuestaSim provides early validation of RTL and the software stacks, particularly in functional verification. Thanks to the co-simulation mode in VPSim, SoC designers have the complete visibility of the IP's activity in the simulator.

7.4.2. *Co-emulation mode*

The co-emulation mode offers a condition close to the real implementation, but with less visibility of the IP's activity. This mode connects VPSim and FPGA in order to simulate the entire system based on SystemC/TLM and RTL implementations. The communication flow in the co-emulation mode is also described in Figure 7.4. Similar to the method in co-simulation, the Remote Target in co-emulation is also connected to an external remote process via a TCP socket. Thanks to the communication using the TCP socket, the connection of VPSim to external prototyping devices is not only limited to local FPGA. The co-emulation mode can also use a remote FPGA connected via Ethernet. The remote process is responsible for forwarding read/write accesses from the TLM initiator to the HW Transactor, through a physical interconnection between CPU and FPGA, such as PCIe. Finally, the HW Transactor implemented on FPGA will perform read and write operations with the design under test (DUT) based on the access requests from the TLM initiator. The Remote Target also supports interrupt handling between VPSim and HW Transactor.

The co-emulation supported by VPSim and FPGA prototypes considers the timing aspect during the system validation. A synchronization mechanism is performed when the virtual and physical prototypes communicate. This synchronization mechanism between VPSim and the FPGA is depicted in Figure 7.4. When using a loosely timed model in SystemC, the simulation of TLM initiators and targets proceeds at every quantum. We use the quantum approach to synchronize VPSim and the FPGA in the proposed framework. The time spent on FPGA, to run peripheral devices, is added to VPSim simulation time at each communication. When a TLM initiator simulated in VPSim sends a read/write request, it calls a b_transport on Remote Target. This call triggers a TCP request to the remote process that runs outside VPSim. The TCP request is then forwarded to FPGA. When the execution on FPGA finishes, it sends an acknowledgment back to the TLM initiator through the same communication path. Afterwards, the FPGA execution will be blocked until the next quantum, where a trigger is sent by the SystemC thread on VPSim. Hence, the execution of VPSim and the FPGA run in parallel with each local quantum, while being synchronized to a global simulation time. For fast functional validation where time accuracy is not necessary, the VPSim and FPGA timing can be decoupled.

7.4.3. *Runtime performance analysis and debugging features*

The proposed hybrid prototyping between VPSim and FPGA offers system validation, which is close to the real execution. To cope with the lack of circuit activity visibility inside the FPGA that hampers debugging capabilities, SESAM provides lightweight monitoring features. They provide sufficient knowledge for efficient analysis, while not incurring a significant penalty to the validation performance. These features are described as follows:

– Performance counters: Performance counters are added to the FPGA to record the number of read/write operations, read/write cycles, read/write bandwidths and access to addresses. This data is collected from the communication buses (AHB and AXI) of the DUT and then summarized to form a statistical report.

– Breakpoint debugger: SoC architects can assign breakpoint addresses on the FPGA in case a thorough debugging process is needed. This can be done through VPSim's interface at runtime. When an assigned address is accessed by a write or read activity, the execution in VPSim and FPGA will be frozen at the exact moment. At this point, we can observe the current performance statistics on FPGA, component states, memory contents, etc.

– Logic analyzer: In HW validation, signal monitoring has proven to be a compelling feature. Circuit monitoring at runtime offers realistic functional and timing validations that can only be provided by a real hardware. For this reason, we add a configurable logic analyzer inside the HW Transactor (see Figure 7.5). Designers can record the selected signals in the FPGA at runtime, for instance, to verify a bug in timing. To minimize the overhead that may reduce the FPGA performance, the recorded signal activities are flushed periodically and saved to a VCD file outside of the FPGA. The beginning and end of signal monitoring are controlled from the VPSim interface.

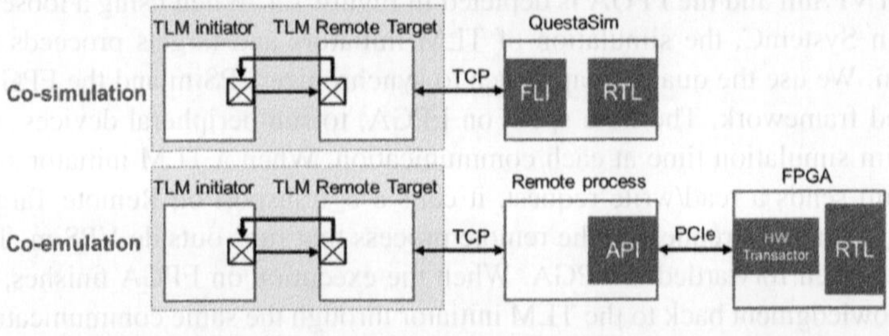

Figure 7.5. *Communication scheme in VPSim during co-simulation and co-emulation*

7.5. FMI for co-simulation

In a CPS, hardware/software components are typically tightly coupled with the physical elements of the system, further increasing its complexity. Therefore, the full cyber–physical context should be taken into account when striving for precise and representative simulation. Simulation can be performed at the system-level, allowing for more potent design tuning and refinement at an early stage. System-level simulation can be performed in different ways. In co-simulation, modeling is done at the subsystem-level. In order to run the system-level simulation, subsystem models are coupled and simulated in a segregated manner.

The functional mock-up interface (FMI) (Blochwitz *et al.* 2011) is the industry standard for multi-domain co-simulation. It defines a modular and interoperable software architecture for co-simulation. It has become popular among industry as well as academia in recent years and is supported by many modeling and simulation tools.

One way of facing the aforementioned challenges is to adapt virtual prototyping solutions for use within system-level simulation environments.

We present here, a fully automated solution for exporting a hardware/software virtual prototype as a functional mock-up unit (FMU). This solution is implemented in VPSim, which is extended in order to support the FMI standard. Therefore, it can easily interface with other FMI-compliant simulators. An FMU encapsulating the virtual platform can be automatically generated based on a high-level description of the hardware/software platform. Also, unlike most of the methods found in the literature, we adopt a more portable and generic approach to time decoupling and parallelism, which is independent of the used FMI master and the hardware models. This work was described in more detail in Saidi *et al.* (2019).

7.5.1. *Functional mock-up interface*

The functional mock-up interface (FMI) is an open standard set in the context of the European ITEA MODELISAR project. The FMI standard was proposed with the goal of facilitating the co-simulation of heterogeneous models. It promotes reusability and exchange of models originating from different modeling tools and defines interfaces to allow their coupling and co-simulation. A model can be exported from a tool that supports FMI as an FMU to be used in FMI co-simulation.

An FMU is a package that contains a combination of files, mainly an XML file listing the variables of the model and the dependency between them. Compiled C code files define functions to instantiate and run the model. Other optional files, such as documentation files, can be included in the package as well. The FMI standard specifies two varieties of FMUs. In FMI for model exchange, the execution tool provides the solver to be used for running the FMU. The functions of the FMU are used to access and compute its variables. In FMI for co-simulation, the solver is provided by the exporting tool and embedded in the FMU.

FMI defines a master-slave architecture for co-simulation. The master algorithm ensures the initialization, setup and orchestration of the co-simulation. It sets the inputs, reads the outputs and controls the advancement of the simulation of each FMU. FMI co-simulation is performed through a calling sequence of FMI C functions. First, an instance of each FMU is created by calling *fmi2Instantiate*. Then, *fmi2SetupExperiment* informs the FMU to set up the experiment, for example, the simulation start and end times. *fmi2EnterInitializationMode* is called to enter the initialization mode of the FMU. This mode is exited by calling *fmi2ExitInitializationMode*. At this stage, the simulation can be run. The FMI master controls the advancement of the simulation by calling *fmi2DoStep*, that runs the simulation for one time step until the end time is reached. During this execution, FMUs exchange data through the FMI master, which calls *fmi2Get* and *fmi2Set* functions to read (resp. write) the outputs (resp. inputs) of each FMU. When the simulation is terminated, *fmi2Terminate* is called. *fmi2Reset* is called to reset the FMU after a simulation run. The FMU is brought back to the same state after *fmi2Instantiate* is called.

7.5.2. VPSim integration in FMI co-simulation

We present a generic and fully automatic method for exporting an FMU, that contains a complete hardware/software virtual platform. The goal is to generate an FMI for co-simulation FMU, i.e. a standalone FMU that can be simply coupled with other FMUs within an FMI co-simulation. Also, we propose a parallel implementation of the generated virtual platform FMU, in order to improve the performance of the co-simulation.

We implemented the FMI interface, as specified by the FMI 2.0 standard, to allow reading the inputs, writing the outputs and controlling the simulation of the virtual platform by the FMI master. We created the *FmiValue* special component in the VPSim library. This component takes care of providing an up-to-date value of one FMI variable to the SystemC subsystem. It is also capable of notifying the SystemC world of value changes, as well as automatically synchronizing with the main thread, as will be explained hereafter. An *FmiValue* can be connected to different types of transport, such as CAN buses, or I2C slaves. This forces the software to effectively implement the low-level communication parts before accessing the actual physical values, thereby increasing the amount of code that can be ported, as is, to the real hardware target. The *FmiValue* component is parameterizable to specify the type of the value, i.e. a real, integer, string or Boolean. Therefore, an *FmiValue* component is created for each variable, that is exchanged between the virtual platform and the physical model.

We propose a user-friendly method for the automatic generation of the virtual platform FMU, where the user has to only make some declarations in the high-level description of the platform. Values that need to be read from or written to the FMI master can be specified within the communication component that transmits them. For instance, if a CAN bus is used for control data transmission between the controller and the vehicle (between the virtual platform FMU and the vehicle dynamics FMU), *FmiValues* components corresponding to the values that need to be transmitted, have to be simply specified inside the CAN component description in the high-level description script of the virtual platform. If at least one *FmiValue* is declared in the platform description, VPSim automatically exports the virtual platform as an FMU, instead of generating a standalone SystemC simulation. Figure 7.6 shows an example of a CAN bus definition taken from the high-level configuration of an automotive SoC. Note that FMI values are declared along with the relevant transport for convenience. Each FMI value is mapped to a unique bus ID, which the software uses to access the corresponding variable.

```
’can’: [
  {
    ’name’: ’RenesasRcarH3CAN’,
    ’base’: 0xe6c30000,
    ’size’: 0x1000,
    ’irq’: 186,

    ’fmi_values’: [
     {
      ’name’: ’speed’,
      ’can_id’: 0x10,
      ’causality’: ’input’,
     },
     {
      ’name’: ’accel’,
      ’can_id’: 0x11,
      ’causality’: ’output’,
     },
     {
      ’name’: ’brake’,
      ’can_id’: 0x12,
      ’causality’: ’output’,
     },
    ],
  },
],
```

Figure 7.6. *Example of* FmiValue *declaration*

Figure 7.7 shows the structure of the generated FMU. In accordance with the FMI standard, it contains a *modelDescription.xml* file, which lists the variables that are exposed to the FMI master. In addition, it contains a set of resource files. The *vplatform.xml* file describes the virtual platform to be simulated. Software binaries, for example, the hardware emulator and the operating system image to be run on the virtual platform, are also included in the FMU. The FMU also contains the C code that is executed on the virtual platform. Finally, the FMU encapsulates VPSim executable, which creates a SystemC simulation of the virtual platform based on the *vplatform.xml* file. This simulation includes *FmiValue* modules which act as proxy components between the virtual platform and the FMI master.

In order to improve the performance of the co-simulation, we propose a parallel implementation of the generated virtual platform FMU. As such, SystemC simulation runs in parallel with the FMI master. The latter is executed within the main thread, hereafter referred to as the FMI thread. When the virtual platform FMU is instantiated by the FMI master, a new thread is created and launched by the FMI thread, in order to run the SystemC simulation of the virtual platform.

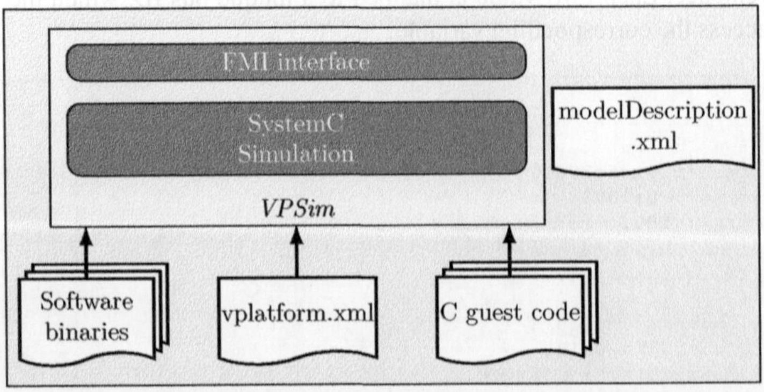

Figure 7.7. *Structure of the generated virtual platform FMU*

By default, the FMI thread and the SystemC thread run independently as long as no variable is exchanged between the two worlds. When one of the two threads accesses a variable, it must make sure that the access does not occur too far ahead in the future from the other thread's viewpoint. Therefore, the accessing thread must wait until the other thread is within a user-defined quantum from its local time. The quantum can be defined on a per-variable basis. This is to account for cases where some variables may be more critical than others and we want to make sure that we do not miss too many updates. *FmiValue* objects can also poll for changes using a user-defined polling period. This can trigger a synchronization with the FMI thread as well.

Figure 7.8 illustrates this parallel implementation and the employed synchronization mechanism. The FMI thread is represented by the black solid arrow, and the SystemC thread is represented by the red solid arrow. In this example, the used quantum is $q = 100\mu s$. When the time step of the FMI co-simulation reaches $500\mu s$, the FMI master proceeds to a variable access. At this instant, the SystemC time stamp is only $300\mu s$. Therefore, the variable access cannot be completed because the time shift Δt is greater than the specified quantum ($\Delta t = 200\mu s > q$). Consequently, the FMI thread waits for the SystemC simulation to progress until it reaches the time stamp $400\mu s$. At this point, the time shift becomes equal to the quantum ($\Delta t = q$) and the FMI thread completes the variable access.

The suggested method for encapsulating the virtual prototyping framework VPSim within an FMU is user-friendly, self-contained and fully automatic, offering a high degree of flexibility. Thanks to the parallel implementation, a good trade-off between the simulation performance and numerical result quality can be obtained.

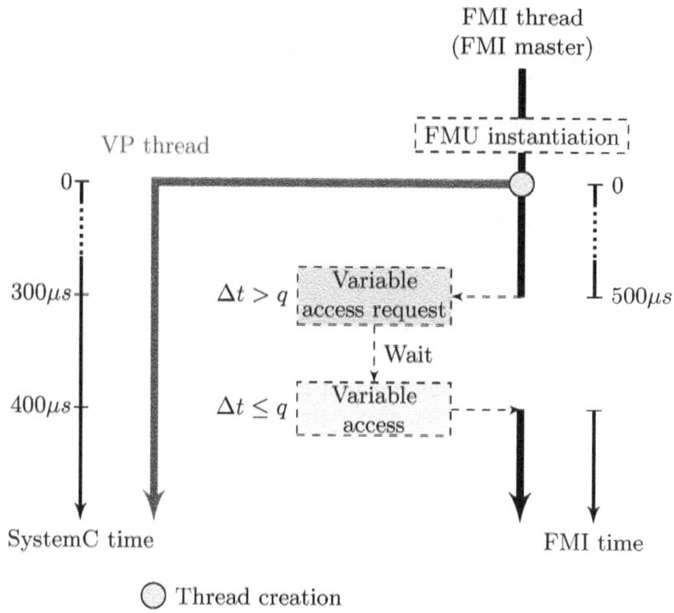

Figure 7.8. *Parallel implementation of the virtual platform FMU*

7.6. Conclusion

This chapter has introduced a complete CPS prototyping solution, based on a set of key contributions. First, we have shown how SESAM effectively enables the fast prototyping of "cyber" subsystems by offering a unified, powerful system description and simulation environment that can seamlessly mix components of various sources (any virtual provider, RTL designs, user devices) into a single prototype. SESAM leverages the latest technologies to accelerate validation and exploration, such as dynamic binary translation, parallelism, time decoupling, etc. We have shown that prototypes built using our framework, were meant to be extended in order to build ever more complex systems. This is made easy by SESAM through the support of standard interfaces, such as SystemC, as well as custom interfaces such as the `Remote` interface, or Python subsystems. To close the loop, we have shown how electronic system prototypes built using SESAM could, with no effort, be co-simulated with continuous and analog subsystems through the FMI standard.

7.7. References

Bellard, F. (2005). QEMU, a fast and portable dynamic translator. *USENIX Annual Technical Conference (ATEC)*, Anaheim, CA, 41–41.

Blochwitz, T., Neidhold, T., Otter, M., Arnold, M., Bausch, C., Monteiro, M., Clauß, C., Wolf, S., Elmqvist, H., Olsson, H., Junghanns, A., Mauss, J., Neumerkel, D., and Peetz, J.-V. (2011). The functional mockup interface for tool independent exchange of simulation models. *Proceedings of the 8th International Modelica Conference*, Linköping Univ. Electronic Press, Dresden, Germany.

Busnot, G., Sassolas, T., Ventroux, N., and Moy, M. (2020). Standard-compliant parallel SystemC simulation of loosely-timed transaction level models. *Proceedings of the 25th Asia and South Pacific Design Automation Conference (ASP-DAC 2020)*, ACM.

Charif, A., Busnot, G., Mameesh, R., Sassolas, T., and Ventroux, N. (2019). Fast virtual prototyping for embedded computing systems design and exploration. *Proceedings of the Rapid Simulation and Performance Evaluation: Methods and Tools*, ACM, 3.

Deschamps, C., Pétrot, F., Burton, M., and Jenn, E. (2019). Gathering memory hierarchy statistics in QEMU. *Design and Verification Conference and Exhibition Europe (DVCON)*.

Kang, S.-H., Yoo, D., and Ha, S. (2016). TQSIM: A fast cycle-approximate processor simulator based on QEMU. *Journal of Systems Architecture*, 66, 33–47.

Open Virtual Platforms (OVP) (2018). Imperas Ltd. [Online]. Available at: http://www.ovpworld.org.

Rodman, N. (2008). ARM fast models-virtual platforms for embedded software development. *Information Quarterly Magazine*, 7(4), 33–36.

Saidi, S.E., Charif, A., Sassolas, T., Le Guay, P.-G., Souza, H.V., and Ventroux, N. (2019). Fast virtual prototyping of cyber-physical systems using SystemC and FMI: ADAS use case. *Proceedings of the 30th International Workshop on Rapid System Prototyping (RSP'19)*, ACM, 43–49.

Sandionigi, C. and Heron, O. (2019). Estimation of oxide breakdown effects by fault injection. *25th Internaltional Symposium on On-Line Testing and Robust System Design (IOLTS)*, 269–274.

Synopsys (2019). Silicon Design & Verification [Online]. Available at: https://www.synopsys.com/verification.html.

Ventroux, N. and Sassolas, T. (2016). A new parallel SystemC kernel leveraging manycore architectures. *Proceedings of the 2016 Conference on Design, Automation & Test in Europe, EDA Consortium*, 487–492.

Wicaksana, A., Charif, A., Andriamisaina, C., and Ventroux, N. (2019). Hybrid prototyping methodology for rapid system validation in HW/SW co-design. *Proceedings of the Conference on Design and Architectures for Signal and Image Processing (DASIP2019)*, IEEE, 35–40.

8

StaccatoLab: A Programming and Execution Model for Large-scale Dataflow Computing

Kees VAN BERKEL

Department of Mathematics and Computer Science,
Eindhoven University of Technology, The Netherlands

Large-scale parallel computing is about organizing millions of arithmetic operations in parallel, *each clock cycle*. Although the state-of-the-art rapidly moves towards GPU-accelerated exascale computing, the typical utilization of the arithmetic resources is only about 2–3%. Using large FFTs as a benchmark, we identify some of the root causes of this underwhelming performance, namely unbalanced hardware architectures and limitations of the applied programming and execution models.

The dataflow programming model is then reviewed as a candidate for large-scale computing. StaccatoLab is introduced as a highly expressive dataflow programming language, supporting synchronous dataflow, cyclo-static dataflow, Boolean dataflow and non-determinate dataflow. StaccatoLab has a well-defined dataflow execution model, designed for efficient and large-scale parallelism. Crucially, this execution model supports *quantitative analysis* of dataflow *programs*, and enables program design for given throughput targets. After a description of the StaccatoLab tools,

For a color version of all figures in this book, see www.iste.co.uk/andrade/multi2.zip.

Multi-Processor System-on-Chip 2 – Applications,
coordinated by Liliana ANDRADE and Frédéric ROUSSEAU.

including Verilog synthesis, this chapter ends with a discussion on the outlook for dataflow programming for large-scale parallel computing.

8.1. Introduction

Parallel computing is an impressive success story. Single-chip performance exceeds 10 TFLOPS (10^{13} floating point operations per second), and the fastest supercomputers have a peak performance of about 100 PFLOPS (10^{17} FLOPS). Assuming a clock frequency of 1 GHz, the latter allows for 10^8 operations per clock cycle. The challenge of parallel *programming* then becomes organizing a given computation such that each of the 10^8 scalar units is busy *every* clock cycle, or to get reasonably close to such 100% utilization.

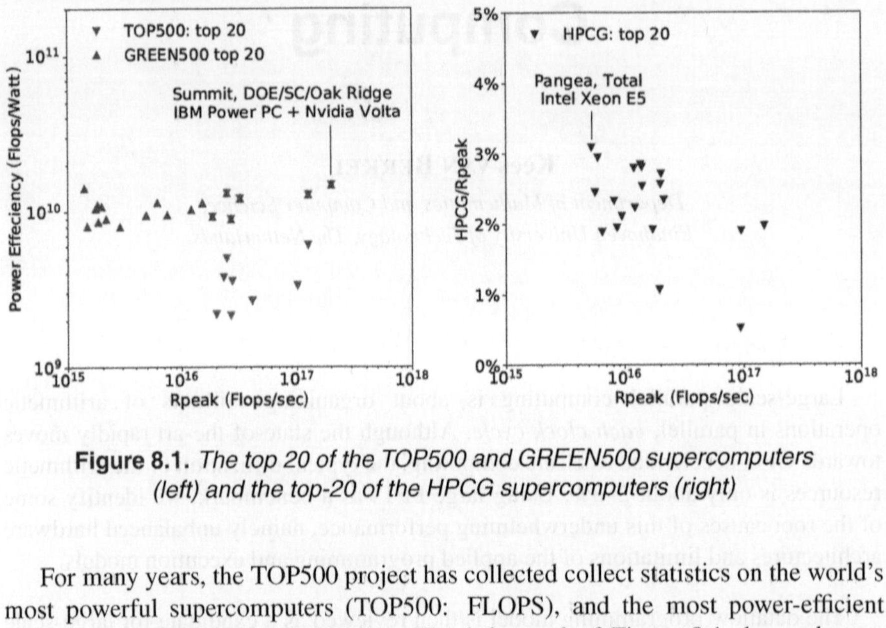

Figure 8.1. *The top 20 of the TOP500 and GREEN500 supercomputers (left) and the top-20 of the HPCG supercomputers (right)*

For many years, the TOP500 project has collected collect statistics on the world's most powerful supercomputers (TOP500: FLOPS), and the most power-efficient computers (GREEN500: FLOPS/W). The left panel of Figure 8.1 shows the top 20 of each, as of November 2019. The state-of-the-art in power efficiency is about 10 GOPS/W. Since 2015, the HPCG (HPCG 2019) benchmark project proposed a new metric for ranking supercomputers. The HPCG (High Performance Conjugate Gradient) "is designed to exercise computational and data access patterns that more closely match a different and broad set of important applications". The right panel of Figure 8.1 shows the top 20 computers in terms of HPCG performance, which is in terms of the HPCG/peak FLOPS ratio. The best HPCG performance is only about 2–3% (!) of its advertized performance.

The published performance of the Summit DOE/SC/Oak Ridge computer, the #1 in the TOP500, is 148.6 PFLOPS. It consumes 10.1 MW, corresponding to 14.7 GOPS/W. The HPCG performance of Summit is 2.9 PFLOPS, and the power efficiency will therefore be closer to 0.3 GOPS/W. What is the story behind these underwhelming numbers? A closer look at different implementations of the FFT (fast Fourier transform) helps to shed some light on this discrepancy.

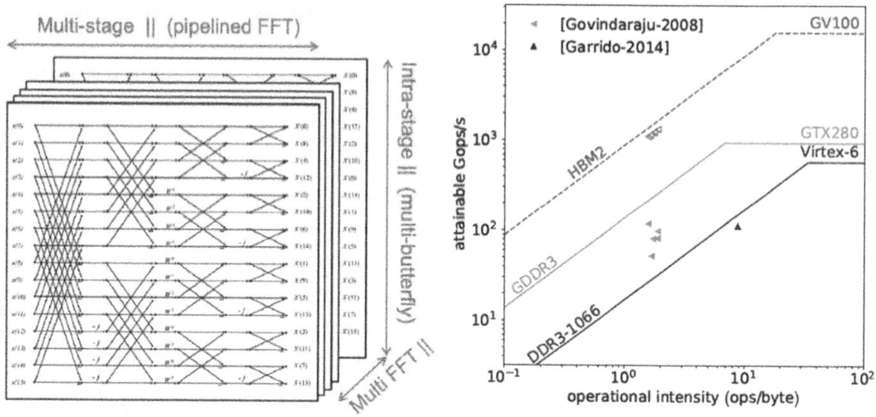

Figure 8.2. *The three axes of FFT parallelism (left) and various machine rooflines with FFT performances*

The FFT is a very common function both in embedded computing and in high-performance computing (HPC). When $N = 2^K$, an N point FFT computation can be organized according to the well-known butterfly graph of the left panel of Figure 8.2. Here, the $N = 16$-point FFT is laid out in $K = 4$ stages of $\frac{N}{2} = 8$ butterflies each. Each butterfly represents 1 complex multiplication and 2 complex additions, or 10 real arithmetic operations (4 mul + 6 add). A complete FFT thus represents $K \cdot \frac{N}{2} \cdot 10 = 5N \log_2 N$ operations. Furthermore, an $(M \times N)$-point FFT can be *Cooley–Tukey* factorized into M N-point FFTs plus N M-point FFTs. This holds for both the 1- and 2-dimensional and versions. Three forms of FFT parallelism can now be exposed (Figure 8.2, left panel):

1) *intra-stage* parallelism: all butterflies within a stage can be computed in parallel. This is raw *data parallelism*;

2) *multi-FFT* parallelism: the M instances of N-point FFTs (needed to compute an $(M \times N)$-point FFT) are independent and can therefore be computed in parallel;

3) *multi-stage* (pipeline) parallelism: butterflies of different stages can be computed in parallel, *provided* the inter-stage data dependencies are respected.

The bulk of the compute power of today's supercomputers is provided by GPUs (graphics processing units). Govindaraju *et al.* 2008 offers a detailed description of how to compute FFTs on such GPUs. GPUs offer abundant data parallelism in the form of *thread parallelism*. Govindaraju maps a butterfly on one thread, or, more precisely, a radix-8 butterfly. ("To maximize the reuse of data read from DRAM [...], it is best to use a large radix R. However, R is limited by the number of registers and the size of the shared memory on the multi-processors [...] We use R=8.") However, on a streaming multi-processor (SM) "hundreds of threads are needed to hide register read-after-write latencies" (Wong *et al.* 2010). With tens of SMs in a GPU, many thousands of threads may be needed. Therefore, multi-FFT parallelism is also required to provide these many threads. Each of these threads requires a multi-kB register file, and almost all SRAM on a GPU is dedicated to these register files.

The reliance of GPUs on multi-FFT parallelism has an important consequence: the intermediate FFT results do not fit in the GPUs' on-chip memory: they have to be stored in the off-chip-DRAM. For a 64k FFT, the 64k complex numbers are stored in, and retrieved from, DRAM *for every three stages* (radix 8). This increases the overall DRAM traffic to sevenfold. The right panel of Figure 8.2 shows a roofline (Williams *et al.* 2009) of the GTX280 GPU, as well as a number of measured GFLOPS numbers for $1k \leq N \leq 32k$, as reported by Govindaraju *et al.* (2008). Because of the $7\times$ increase in the DRAM traffic, the so-called *operational intensity* numbers end up below 2.

The Summit supercomputer uses over 26000 Nvidia GV100 GPUs of 15.7 TFLOPs each (32 bit). Each GPU has an HBM2 DRAM interface with a whopping bandwidth of 900 GB/s! The resulting roofline for the GV100 allows us to (generously) extrapolate the FFT numbers. Bottom line: the FFT FLOPS numbers for the GPU used in the Summit supercomputer are well below 10% of the advertized numbers. This is because the FFT is *memory-bound* on a machine that devotes almost all its SRAM to register files for many thousands of threads that are idle most of the time (>99%).

So-called *pipelined FFTs* exploit parallelism across stages. In its basic form, such a pipeline accepts one complex sample each clock cycle, resulting in a sample throughput equal to the circuit's clock frequency, and a latency of N clock periods. The hardware pipeline has a depth of $\log_2 N$ stages, where each stage has an exponentially decreasing (or increasing) buffer to store $\frac{N}{2}, \frac{N}{4}, \frac{N}{8}, \frac{N}{16}, \ldots$ intermediate results. Crucially, there is no need to access DRAM for intermediate results. The FFTs of Garrido *et al.* (2013, 2014) combine pipelining with intra-stage parallelism, so that the FFT can accept, for example, 16 samples each clock cycle. As a result, the 16-wide 64k-point FFT achieves a throughput of 2.29 Gsamples/sec on a 143 MHz FPGA (Figure 8.2). The pipelined FFTs are synthesized from VHDL. Compared to the GPU FFT, a 5-10$\times$ higher operational intensity is achieved, and hence a 5-10$\times$ higher performance for the same DRAM bandwidth. For more details, see van Berkel (2016).

Table 8.1 summarizes the FFT exercise. (It is *not* intended as a full benchmark GPU vs. FPGA.) The example of large FFTs highlights some of the limitations of 1) multi-threading as a programming model and 2) the use of excessive thread parallelism to hide DRAM-access latencies as an execution model. On the contrary, mapping an FFT onto an FPGA, such that it has a throughput of 16 samples per clock cycle, amounts to VHDL hardware design.

	FFT on GPU	FFT on FPGA
Parallelism	Data parallelism	Pipelined (+data) parallelism
Programming model	Many thousands of threads	$\log_2(N)$ hardware modules
Programming language	CUDA, OpenCL, etc.	VHDL, Verilog
Synchronization	Coarse ("barrier")	Fine (per clock, between modules)
Communication	Shared memory	FIFO queues
SRAM use	Reg files, cache	Data queues
Operational intensity ($N = 2^{16}$)	1.4	10

Table 8.1. *A comparison of large FFTs mapped onto a GPU and an FPGA*

In this chapter, we explore dataflow as an alternative programming model (sections 8.2 and 8.3) using small but non-trivial examples. The StaccatoLab execution model (section 8.4) supports fine-grained (clock-cycle level) synchronization between processes, and minimizes memory usage. Moreover, it is simple and fully transparent, enabling quantitative analysis of dataflow programs. The tools based on this execution model, as reviewed in section 8.5, include dataflow graphs {simulation, analysis, transformation and mapping to FPGA/Verilog}. Section 8.6 reflects on the perspective of using dataflow programming for large-scale parallel applications.

8.2. Static dataflow

This section, together with the next, offers an introduction of dataflow as a programming model for parallel computations. It does so by means of a sequence of examples that introduce dataflow concepts, as well as describing the dataflow programs, showing results of *timed* simulations. It also demonstrates graph-based performance analysis. All dataflow programs and simulation results can be found in van Berkel (2020). A more technical description of the execution model is deferred to section 8.4.

An early paper on dataflow as a programming model and language is Dennis (1974). It introduces dataflow "in terms of a semantic model that permits concurrent execution of non-interfering program parts", and much of the then introduced dataflow terminology is adopted in this section. Interestingly, the basis foundations of dataflow performance analysis was already laid in the 1960s by Karp and Miller (1966) and Reiter (1968), among others, without using the word dataflow. This chapter also benefits from over two decades of dataflow work at Berkeley University, as consolidated in Ptolemaeus (2014). For a broader and partially historical overview, the reader is referred to Johnston *et al.* (2004).

The basis of any dataflow program is a directed graph, a set of nodes connected by edges.[1] Nodes consume tokens from input edges and produce tokens on output edges. Each token may carry a value of some type, such as a boolean, an integer, a complex number or a fixed vector of 16 complex numbers. Each token value is computed by the producing node. The production and consumption of tokens is governed by so-called *firing rules*. Unlike many dataflow programming models, StaccatoLab assumes finite edge capacities: a firing rule is *enabled* when for each input edge, a specified number of tokens is present *and* for each output edge, a specified number of vacancies is present. A vacancy can also be seen as a token on an implicit back edge. More about that will be discussed later. The number of vacancies is also referred to as *slack*.

Dataflow graphs can be classified by the number of tokens involved in firing rules, the number of different firing rules per node, how the next firing rule is selected for execution and whether multiple firing rules can be enabled at any point in time.

8.2.1. Synchronous dataflow

In a synchronous dataflow (SDF) graph (Lee and Messerschmitt 1987; Lee and Parks 1995), each node has one fixed firing rule. This rule specifies per input a number of tokens to be consumed, and per output a number of vacancies to be filled. *Homogeneous dataflow*, also referred to as single-rate dataflow, is a special case of SDF. In a homogeneous dataflow graph, a node is *enabled for execution* when one token is present at each input and one vacancy present at each output. The execution of a node amounts to the consumption of one token from each input and the production of one on each output.

Graph `G`, described in Figure 8.3 using the Python syntax, is a very simple homogeneous flow graph comprising seven nodes and seven edges. Nodes `n[1]` and `n[5]` fork-join two parallel paths of different lengths. By default, each edge has a capacity of one, has no initial tokens and hence has one vacancy initially. The default execution time for each node equals one time unit, called a (clock) cycle. The graph is depicted in Figure 8.4.

When node `n[1]` fires, it produces two tokens, one on each output. The bottom one takes three cycles to arrive at `n[5]`, while the top one is waiting to join at `n[5]`. When

1 Dataflow nodes are also referred to as processing units, processes, or actors; and edges as arcs, branches, (FIFO) channels, or queues.

each node fires as soon as its firing rule is enabled, a so-called self-timed execution, the flow pattern of Figure 8.4 (bottom left) emerges. After a prelude of three cycles, a periodic behavior of four cycles sets in. Such a pattern is known as a $(3, 4)$ *lasso* (Ghamarian *et al.* 2006). Once periodic behavior has set in, each node fires once every four cycles. The nodes fire at *rate* $= \frac{1}{4}$. The throughput of graph G can be increased by initializing the graph with, for example, three tokens on edge `e[1]`. This leads to a $(4, 1)$ lasso, and a sustained firing *rate* $= 1$.

```
G   = Graph()
G.n = [Node() for i in range(7)]
G.e = Path (G.n)
G.f = Edge (G.n[1], G.n[5])
G.e[1].init(D=3)                  # used only for the second simulation
```

Figure 8.3. *Homogeneous flow graph* `G` *comprising seven nodes and seven edges*

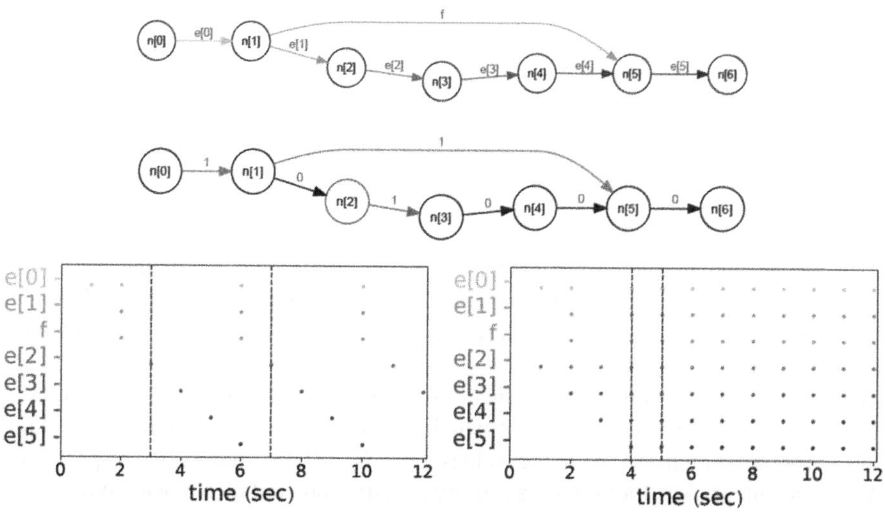

Figure 8.4. *Flow graph* `G` *(top), its state after three cycles (mid), its flow (event timing) during 12 cycles (left) and its flow with three initial tokens on edge* `e[1]` *(right)*

The Lorenz attractor is a dynamical system of three non-linear differential equations [8.1]:

$$\frac{\mathrm{d}x}{\mathrm{d}t} = \sigma(y - x), \qquad \frac{\mathrm{d}y}{\mathrm{d}t} = x(\rho - z) - y, \qquad \frac{\mathrm{d}z}{\mathrm{d}t} = xy - \beta z. \qquad [8.1]$$

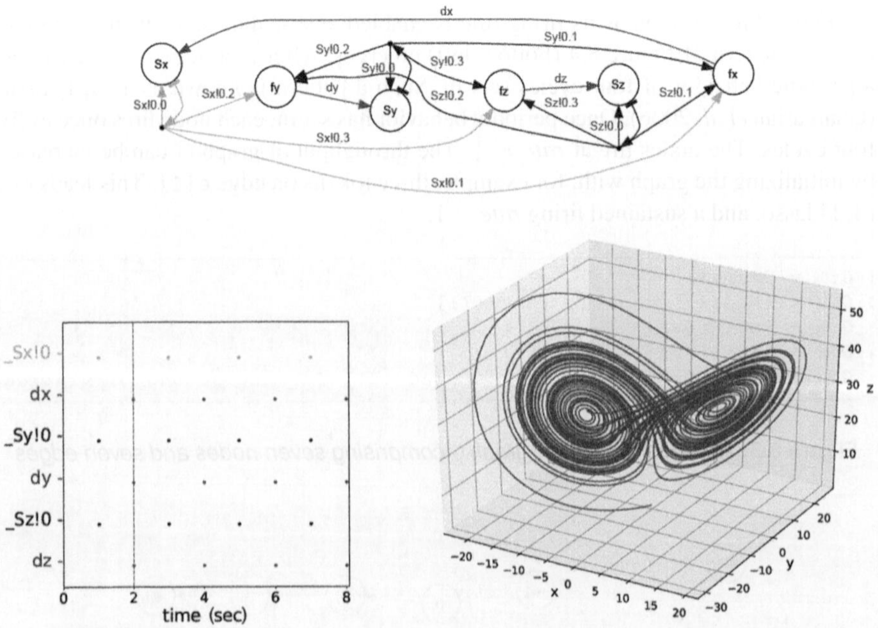

Figure 8.5. *Graph* `Glorenz` *(top), its flow during the first eight cycles (left) and the* `x, y, z` *data produced by nodes* `fx, fy, fz`*, respectively (right)*

With dataflow graph `Glorenz` (Figure 8.5), we aim to numerically integrate these equations over some time interval using Euler integration. The program shown in Figure 8.6 introduces the constants `sigma, rho`, and `beta`, the graph topology with integration nodes `Sx, Sy`, and `Sz` (`dt`= 0.01) and function nodes `fx, fy`, and `fz`, each referring to one of the output functions `fo`. Note that the outputs of each of {`Sx, Sy, Sz`} fork towards four edges, of which one loops back to the node. These three self-loops are initialized with the values `-0.15, -0.2, and 0.2`, respectively. Also, the integration loops {`G.dx, G.dy, G.dz`} are initialized with one token, valued at 0. Simulation shows that `Glorenz` has lasso $(0, 2)$: it takes two cycles for the tokens to traverse the loop. With a few additional lines of Python code, the outputs can be plotted in a 3D cube yielding the well-known Lorenz butterfly.

In graph `Glorenz`, the value of an output token is a function of the values of the tokens consumed in the same firing. In order to support functions that depend on the values of (multiple) past tokens, a *fixed-length input buffer* (size N) can be introduced to an input. With the consumption of a new token, the token value accepted N consumptions ago is discarded. An output function can access each of the N values of this buffer. Such a buffer can be seen as a fixed-length extension of the token queue associated with the input edge, to store the values of previously consumed tokens.

```
sigma,rho,beta, dt= 10, 28, 8/3. 0.01
fo              = [lambda x: sigma*(x[1] - x[0]),
                   lambda x: (rho-x[2])*x[0] - x[1],
                   lambda x: x[0]*x[1] - beta*x[2]]

Glorenz         = Graph(); G=Glorenz
G.Sx,G.Sy,G.Sz= [Node(I=0, fo= lambda x: x[0]+dt*x[1])
                   for i in range(3)]
G.fx,G.fy,G.fz= [Node(I=[G.Sx.O[0],G.Sy.O[0],G.Sz.O[0]],fo=fo[i])
                   for i in range(3)]
G.dx,G.dy,G.dz= [Edge(G.fx,G.Sx),Edge(G.fy,G.Sy),Edge(G.fz,G.Sz)]

for i,n in enumerate([G.Sx, G.Sy, G.Sz]):
    n.I[0].init(D=1, S=0, x=[-0.15, -0.2, 0.2][i])
for e in (G.dx, G.dy, G.dz): e.init(D=1, S=0, x=0)
```

Figure 8.6. *Definition graph* `Glorenz`

```
N     = 11
GMA   = Graph(rate=50e3);  G= GMA
G.mp3 = MP3(mp3='../../../audio/war.mp3', tmin=37.5, rate=44100)
fo    = lambda x: np.int16((x[0][N]-x[0][0])/N +x[1])
G.ma  = Node(I=[G.mp3,0], fo=fo)
G.ma.I[0].init(x=[0 for i in range(N)])
G.ma.I[1].init(D=1, x=[0])          # accumulator, initialized to 0
G.snk = Node(I=G.ma.O[0])
```

Figure 8.7. *Definition of graph* `GMA`

In graph `GMA`, defined in Figure 8.7, input `ma.I[0]` has such a buffer, initialized with N zeros. Output function `fo` computes the difference of the most recent buffer value `x[0][N]` and the oldest value `x[0][0]`. By weighing this difference by $1/N$ and adding it to accumulator `x[1]`, node `ma` computes the *moving average* of the input stream `G.mp3!0`.

This input stream is produced by library node `mp3`. Node `mp3` has two self-loops: one edge stores a *macro-token* containing a constant array of audio samples, and the other edge stores a pointer to the next audio sample to be the output. The *rate* of node `mp3` has been set to 44.1 kHz, whereas the graph rate is 50 kHz. The firing of node `mp3` is therefore *throttled*, causing incidental gaps in the token flow of Figure 8.8. The moving-average function is effectively a low-pass filter, with signal smoothening visible in both the time domain (Figure 8.8, left panel) and the frequency domain (right panel).

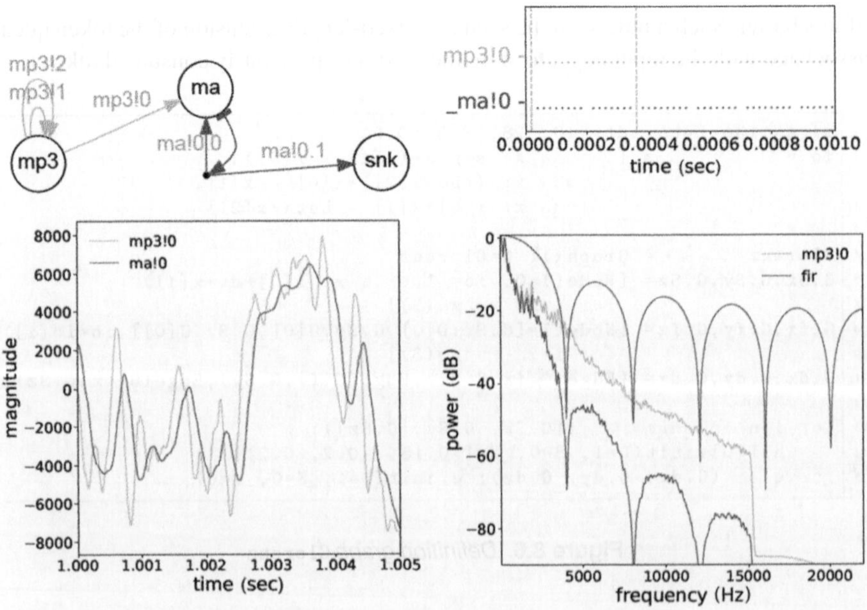

Figure 8.8. *Node* ma *computes the moving average over a window* $N = 11$ *samples. The flow shows incidental gaps due to flow throttling by the* mp3 *source. The two bottom plots show the signal smoothening in time and frequency domain*

8.2.2. *Cyclo-static dataflow*

So far, dataflow nodes have a single (SDF) firing rule. Bilsen *et al.* (1995) introduced *cyclo-static dataflow* (CSDF) to "support algorithms with a cyclically changing, but predefined, behavior". A CSDF node periodically traverses a fixed sequence of states, where each state has its own firing rule. The paper also presents an algorithm to convert a CSDF graph into an SDF graph, as well as a speech-processing application.

```
set_fsm([Repeat(L, iter=Rule(I=[1,0,1], O=[0,1,1], s=0), exit=1),
         Fire(Rule(I=[1,1,1], O=[1,1,1], s=1))])
```

Figure 8.9. *An example of an FSM for a cyclo-static dataflow node*

CSDF node behavior can be specified by a finite state machine (FSM), as shown in Figure 8.9. This FSM example is for a node with three inputs and three outputs. From its initial state `s=0`, a first `Rule` is repeated for `L` firings, before `exit`ing to state `1`. Once in state `1`, a second `Rule` is applied indefinitely. (Technically, this is a minor but significant extension to CSDF, as the FSM never returns to its initial state.) Each `Rule` specifies how many tokens are consumed from each input (the `I` lists), and how many vacancies are required on each output (the `O` lists). Using `Repeat` and `Fire` constructs, FSMs for CSDF nodes can be programmed.

The three nodes `ro[i]` of subgraph `RO` in Figure 8.10 use the above FSM to re-order incoming sample blocks of size N, according to the FFT *bit-reversal* scheme for $N = 64$. Each `ro[i]` starts with loading the first `L` (L_i) values in a queue (self loop), before it continues with appropriately interleaving further inputs with the values in this queue, using an additional modulo-N_i counter (Garrido *et al.* 2011). A cascade of two instances of subgraph `RO` performs the identity function. The resulting graph has a $(105, 1)$-lasso and hence a throughput of 1.

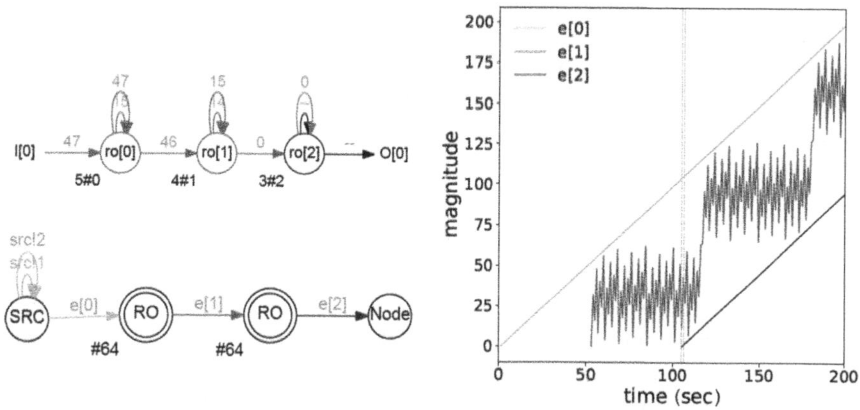

Figure 8.10. *Subgraph* `RO` *(left, top) comprises three CSDF nodes* `ro[i]` *and performs an* $N = 64$ *FFT bit-reversal. As FFT bit-reversal is its own inverse, a cascade of two* `RO`*s performs the identity, as shown for a linear input sequence (right)*

8.2.3. *Dataflow graph transformations*

Parhi and Messerschmitt (1991) introduced J-*unfolding* as a graph transformation that increases the token throughput of a homogeneous dataflow graph J-fold, at the expense of, at most, a J-fold increase in compute resources. The graph transformation increases both the node count and the edge count J-fold. For acyclic dataflow graphs, J may be arbitrarily large. However, for graphs with cycles (including self-loops), the throughput gain is bounded by the initial token counts.

In StaccatoLab, graph-unfolding is implemented at the FSM level, thereby generalizing unfolding to CSDF graphs. In a static dataflow graph, each node has the property that, in a given state, only one firing rule is available. Only when that rule is enabled, is a next firing possible. For any CSDF node, the firing rules can be put in a sequence that, after a prelude, is periodic. The J-unfolded FSM is constructed by grouping J successive firing rules into new J-unfolded rules. When P is the length of the period of the original FSM, the period of the J-unfolded FSM becomes the least common multiple of P and J. FSM unfolding does not change the topology of the graph, but typically requires an increase in edge capacities. In StaccatoLab, J-unfolding is automated, but may rely on the manual unrolling of tight graph cycles.

The J-unfolding of a homogeneous (sub)graph results in a dataflow (sub)graph where each node consumes J tokens and produces J tokens each cycle. Obviously, this requires a J-fold increase in the resources required for evaluating the node output function `fo()`. This is not unlike SIMD (single instruction multiple data) parallelism.

The goal of the *K-slow transformation* (Parhi 2007) is to share a homogeneous subgraph among K clients. This sharing amounts to a token-by-token interleaving of K input streams before offering it to the shared subgraph, and de-interleaving the streams afterwards. Typical applications include stereo audio streams and RGB pixel streams. The K-slow transformation is achieved by replacing each initial token by K copies of that token. Examples of J-unfolding and K-slow transformations applied to StaccatoLab programs can be found in van Berkel (2018).

8.3. Dynamic dataflow

In a static dataflow graph, every node is static, i.e. each node state has a unique successor state with only one firing rule to make the transition. Hence, the sequence of firing rules of a static node has the form of a lasso, similar to the self-timed firing behavior of the SDF graph. A node is *dynamic* if it has a state with multiple alternative successor states and multiple firing rules leading to these states. A dataflow graph is dynamic if it contains at least one dynamic node.

The selection among multiple firing rules leading to one of the multiple next states can be based on:

1) the value(s) of incoming token(s), also known as *data-dependent* dataflow;

2) the absence/presence of input tokens, also known as *non-determinate* dataflow.

Both forms of dynamic dataflow can be much harder to analyze. For example, whether or not a dynamic dataflow graph can be implemented with finite buffer sizes is *undecidable* in general (Buck and Lee 1993). Nevertheless, the additional expressive power of dynamic dataflow is essential for many practical parallel programs.

8.3.1. *Data-dependent dataflow*

Boolean dataflow (BDF) is the quintessential form of data-dependent dataflow. BDF only allows two types of dynamic nodes: SWITCH and SELECT, as shown in

Figure 8.11 (Dennis 1974). Both have a so-called control port (labeled c) that accepts boolean-valued tokens. The value of the incoming control token specifies which output port the incoming data token must be forwarded to (SWITCH) or which input port the token to be forwarded must be selected from (SELECT).

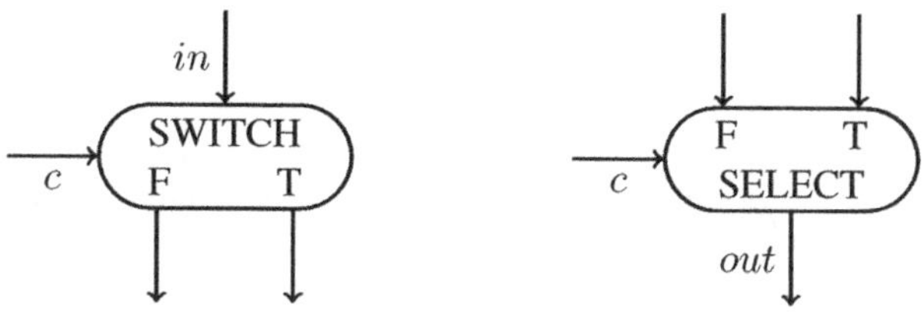

Figure 8.11. *SWITCH and SELECT nodes as used in Boolean Dataflow*

Lee (1991) defined a dataflow graph as *consistent* “if on each edge, in the long run, the same number of tokens are consumed as produced”; Lee also introduced some calculus of proportions (not probabilities!) of specific control-token values, applying calculus to proportions to analyze consistency, and offered ample examples of (in)consistent graphs. Buck and Lee (1993) showed that BDF is Turing complete, meaning that for any problem that can be solved by a Turing machine, there exists a BDF graph that does the same. This fact also implies that proving the consistency of a BDF graph is undecidable.

There are many proposals to constrain dynamic dataflow such that some form of consistency can be verified, essentially by constraining the state space. For example, Gao *et al.* (1992) introduced *well-behaved dataflow*, where the use of SELECT and SWITCH nodes was limited to either a *conditional schema* (like if-then-else) or a simple *loop schema*. These syntactically constrained forms of dynamic dataflow have the advantage that the boundedness of buffers can be decided at compile time. The StaccatoLab programming model follows a somewhat different approach. First, it generalizes BDF by means of so-called *select functions* to achieve significantly more compact programs. Second, it assumes that proving consistency (and deriving buffer bounds) is the responsibility of the programmer and is not that difficult for *specific* programs.

`Sort.64` is a recursive dataflow graph comprising an instance of `Sort.32`, node `spl` (split) and node `mrg` (merge); see Figure 8.12. The idea is that `Sort.32` outputs blocks of 32 integer-valued tokens, where each block is sorted in increasing order. Node `spl` forwards a sorted block alternatingly along output 0 and along output 1. Node `mrg` merges the two sorted blocks into a sorted block of double length. A token value `x` is a tuple, with `x[0]` $\in \{0, 1\}$, where `x[0]=1` marks the end (last token) of a block. These markers enable block-size independent FSMs for `spl` and `mrg`.

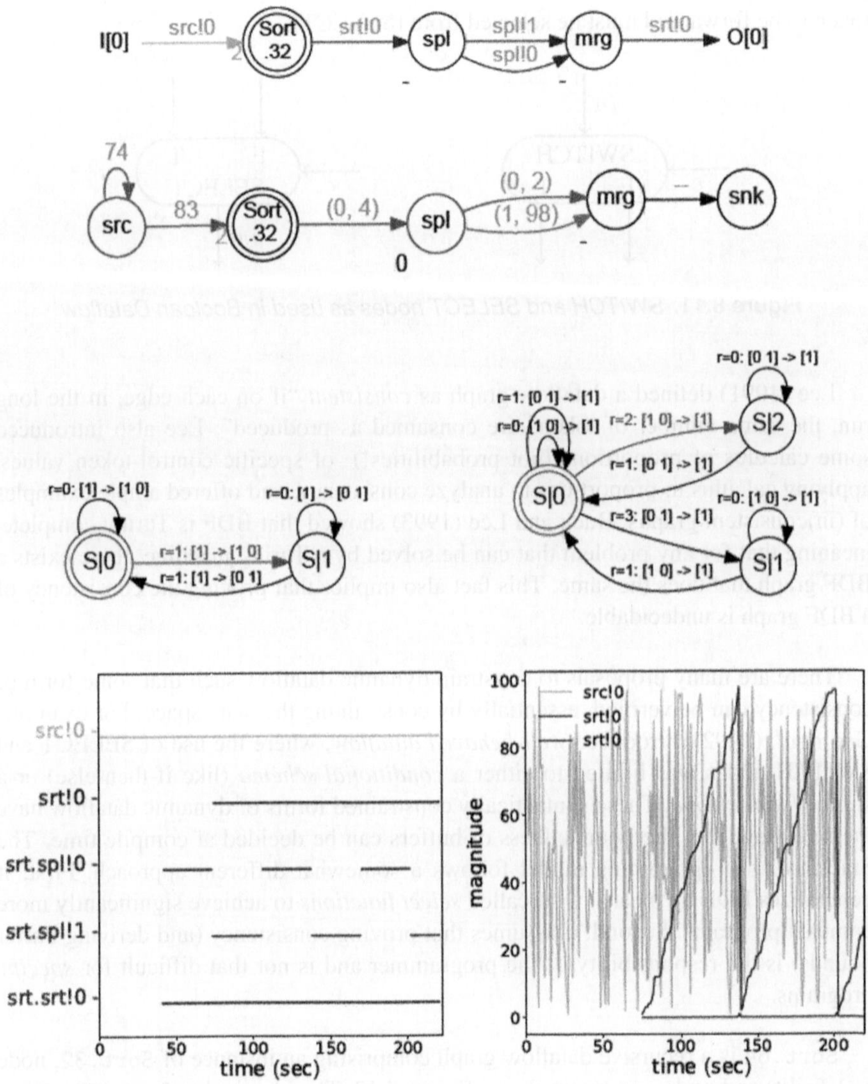

Figure 8.12. *Recursive dataflow graph* `Sort.64` *(top),* `Sort.64` *connected to a random-number source* `src` *and* `snk` *(one-from-top), the FSMs of nodes* `spl` *and* `mrg` *(mid-left and right) and the flow and data plots for* `Sort.64` *(bottom)*

Node `spl` is an instance of *node class* `Split`. See the program in Figure 8.13 and graph in Figure 8.12, mid-left. For both FSM states, numbered 0 and 1, a `Select` specifies two possible firing rules. The selection between these two rules is governed by a *select function* `fs`: the value returned by `fs`, here the value of the token marker, determines which of the two `Select` rules to be executed, and hence the next state. The net effect is that a `Split` node alternately forwards a block to one output and then to the other.

```
class Split(Node):
    def __init__(self, N, I=[]):
        super(Split, self).__init__(I=I)
        pass0 = lambda s: Rule(I=(1,), O=(1,0), s=s)
        pass1 = lambda s: Rule(I=(1,), O=(0,1), s=s)
        self.set_fsm([Select([pass0(0), pass0(1)]),
                      Select([pass1(1), pass1(0)])])
        self.set_fs([lambda x: x[0], lambda x: x[0]])
        self.set_fo([lambda x: x,    lambda x: x])
```

Figure 8.13. *Definition of node class* `Split`

Node `mrg` is an instance of node class `Merge` with a three-state FSM (Figure 8.12, mid-right). In state 0, it forwards the smaller-valued token at its input, until it has forwarded an end-of-block token. Then, it transits to either state 1 or 2, so that the uncompleted block can be depleted, regardless of what happens at the other input. When the second input block has also depleted, the node returns to its initial state.

Rather than analyzing graph `Sort.64` for given queue sizes, it is more interesting to address the question: *What edge capacities are needed to sustain throughput = 1?* The recursive graph begs for an argument by induction. Let's assume that `Sort.32` offers a next value each cycle. Node `spl` can forward the first block when `spl!0` can hold at least 32 tokens. Node `mrg` can fire a first time if it has a token on each input. If the first `spl!1` token value exceeds all 32 token values on `spl!1`, queue `spl!0` is depleted first, and `spl!1` must also have a capacity of at least 32. However, in the other extreme case, queue `spl!1` is depleted first, and *the first token of the next 32-block* arrives at the same cycle as the last token is removed from `spl!1`. Hence, the capacity of `spl!0` must exceed 32+1. (A node `Sort.2` with throughput=1 is straightforward.)

Figure 8.12 (bottom panels) shows a flow and data plot for random inputs. Interestingly, the capacity of `spl!1` can be reduced down to 1, with the worst case throughput being reduced to 0.5. For a sorter of block size N, the required on-chip memory equals $2N$. On-chip memory may not suffice for very large blocks. Then, a multi-pass dataflow graph is needed which makes use of an off-chip DRAM to store sorted sub-blocks. A sorter for arbitrary block sizes $\leq N$ is a bit more involved.

8.3.2. *Non-determinate dataflow*

The quintessential *non-determinate* dataflow node is the *non-determinate merge* (not to be confused with the merge sorter above), as introduced by Arvind and Brock (1984). Lee (2006) describes this node as "an unordered merge where token streams can be merged in unpredictable ways, depending on the time of arrival of tokens". A binary (two-input) merge has two firing rules: forward from input 0 and forward from input 1. The selection between the two depends on the availability of tokens. When tokens are available on both inputs, the choice between the two rules is non-deterministic, i.e. the programmer cannot know which one. In general, dataflow graphs with non-determinate nodes can be very hard to analyze, but well-contained forms of non-determinate dataflow, such as the merge, are relatively benign.

```
class IMG(Node):
    def __init__(self, N I=[]):
        super(IMG, self).__init__(I=I)
        rules= [Rule(I=[int(j==i) for j in range(N)],O=[1,],s=0)
                  for i in range(N)]
        self.set_fsm([Choice(rules)])
        self.set_fo([lambda x, r: x[r]])
```

Figure 8.14. *Definition of node class* `IMG`

As shown in Figure 8.14, a node of class `IMG` ("input merge") is an `N`-way merge with `N` inputs and one output. Python variable `rules` defines a list of `N` firing rules `Rule(..)`, where rule `i` consumes one token from input `i` and produces one output token. The `Choice` constructor applied to `rules` defines the node's FSM. The index `r` of the chosen rule is used as an argument of the output function.

Dataflow graph `Farm` (Figure 8.15) is a toy processor farm. Node `img` is an instance of input merge `IMG(3)`. Node `omg` is a three-way "output merge": it *de*merges token stream `src!0` into streams `omg!`i, for $0 \leq i < 3$. Node `omg` forwards a token to an output that has a vacancy. If there is no vacancy, the choice among the three firing rules is postponed until there is a vacancy. Nodes `p[`i`]` are dynamic nodes: upon reception of a token, they cycle the token a random number of times (≤ 100) on their self-loops before forwarding the original token value. The random timing of the processor nodes is clearly visible in the flow plot, and the data plot shows that the output tokens are merged out-of-order.

The StaccatoLab (library) node `RAM` is another example of a non-determinate node: it merges streams of read and write transactions. A read transaction is a firing rule that inputs a read-address token and outputs the `RAM` content at that address. A write transaction is a firing rule that accepts two tokens: a write-address and a value to be written. The content of the `RAM` is an array stored as a macro-token on its self-loop.

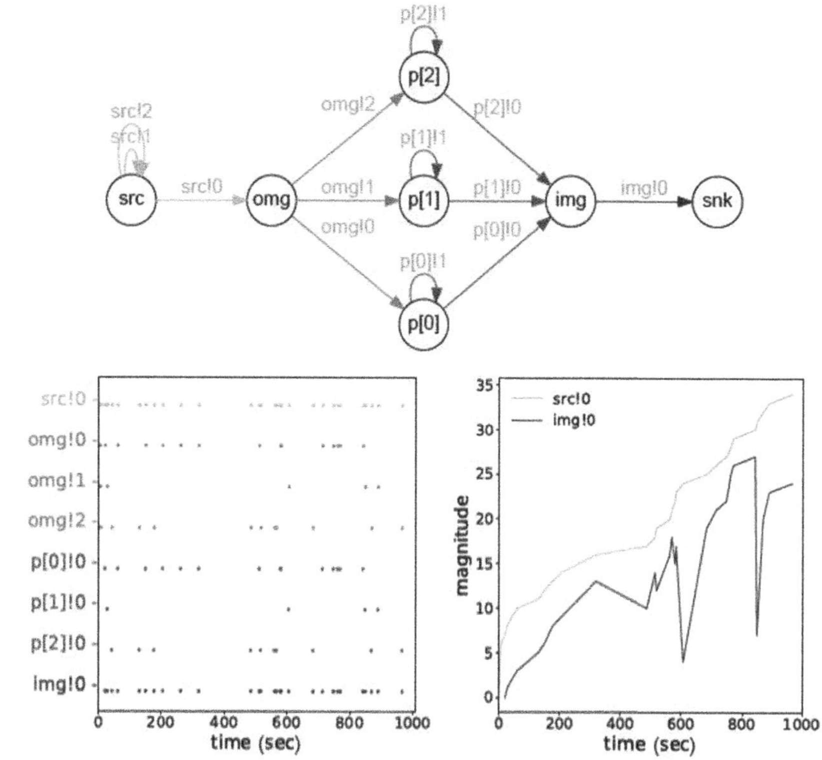

Figure 8.15. *A toy processor farm: dataflow graph (top), flow plot (left) and data plot (right). Node src produces the natural numbers, in order*

Dataflow graph `Sobel` (Figure 8.16) performs a Sobel filter (magnitude only) on an image stored in `RAM` node `ram` and stores the result in a different segment of the `ram` memory. During a single `ram` read transaction, a burst of $B = 16$ consecutive pixels are read. Node `unp` unpacks this B-tuple and node `pck` does the complementary packing before the filtered B-tuple is written back to memory. Both are SDF nodes: `unp` accepts one token and produces B tokens, `pck` accepts W and produces one.

Nodes `dl0` and `dl1` implement *delay lines*, or, more accurately, their input edges store W pixels with W the image width. Nodes `sbx` and `sby` implement the two 3×3 convolution kernels and `mag` computes the magnitude of the output pixel. The latter five nodes are all homogeneous nodes. The flow plot in Figure 8.16 clearly shows throughput=1 for the filter function. The `ram` accesses are nicely merged, and only 2/W `ram`-access cycles are thus used. The unused memory bandwidth can be used by applying J-unfolding (automated) to the filter nodes, up to $J = 8$. Each filter node then becomes simple J-SIMD processors. Note that the memory costs of the delay lines are not affected by this J-unfolding.

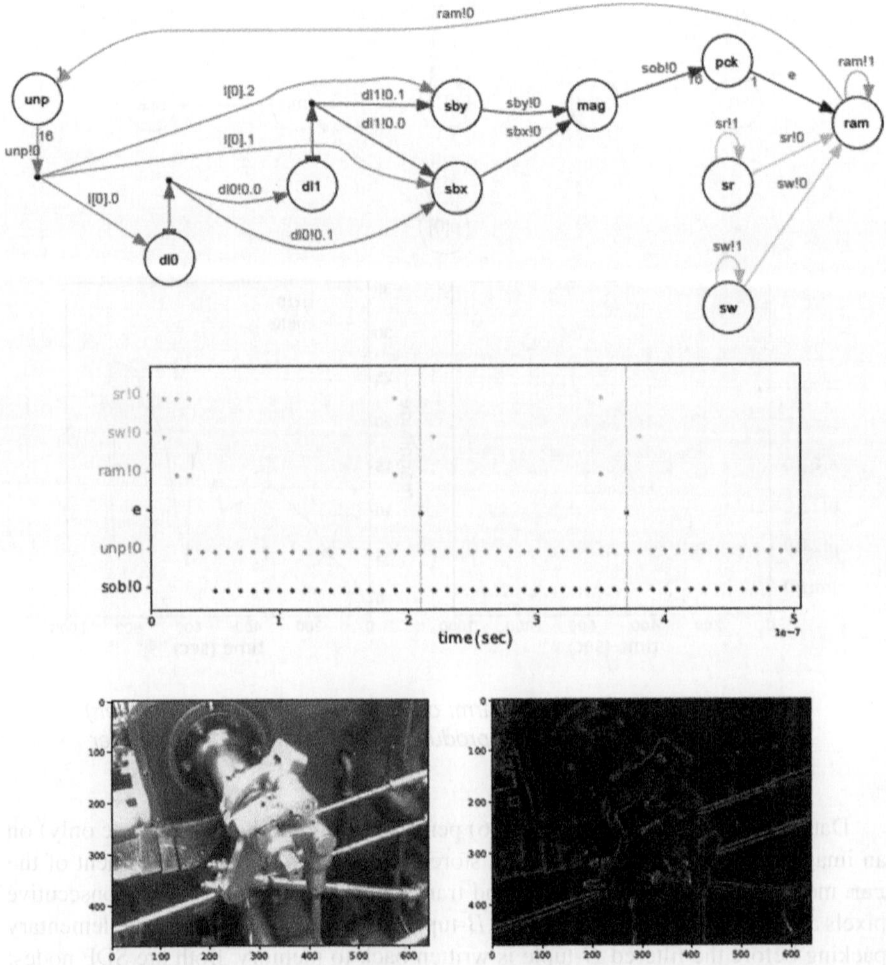

Figure 8.16. *Dataflow graph* `Sobel` *with I/O to* `RAM` *node* `ram` *(top), 50μsec flow plot (middle) and example input+output images (bottom)*

StaccatoLab supports the non-determinate choice among arbitrary sets of firing rules, although practical applications are about (de)merging data/transaction streams. The default arbitration is round-robin, with priority-based as an option. So, what then is the issue about non-determinate behavior? When a dataflow graph is mapped onto an FPGA, an edge with a capacity of say, 1 MB, may end-up in an off-chip DRAM. With careful design, flow-rates can still be made predictable. However, the DRAM controller may introduce variable token delays, possibly affecting the *schedules* of non-determinate nodes. In most practical designs, these delays must not affect the *function* of the dataflow graph. A similar concern is known as memory consistency which is about avoiding write-after-read (WAR) hazards.

8.4. Dataflow execution models

Dataflow as a programming model specifies *what* is computed. A dataflow execution model is needed to specify *how* the computation is organized and must provide answers to questions such as: How much parallelism? At what granularity? How is all the work scheduled? What synchronization primitives are used? How are events timed? How is memory organized? For example, a possible execution model for dataflow graphs can be based on a fully sequential schedule, so that it can run on a single CPU. This execution model is used to simulate dataflow graphs.

The StaccatoLab execution model has the following objectives. It must:

1) offer maximum parallelism and hence maximum throughput;

2) scale to arbitrarily large graphs;

3) be simple and transparent;

4) ...and thus quantitatively enable reasoning about program performance;

5) allow an efficient implementation on (multi-)FPGA platforms.

The execution model will be described in terms of the basic dataflow theory and in part by augmenting the dataflow graph. The execution model will be explained for the simpler case of homogeneous dataflow. Most model features extend to the more general forms of static and dynamic dataflows in a fairly straightforward fashion.

8.4.1. *A brief review of dataflow theory*

For the formal underpinning of dataflow, we follow Reiter (1968) and Lee and Parks (1995). For the time being, we focus on homogeneous dataflow: with each node, firing 1 token is consumed from each input and one token is produced on each output. A graph is *consistent* if, in the long run, production and consumption are in balance on each edge token. Only for a consistent graph is an implementation with bounded queues feasible. So-called balance equations and elementary linear algebra can be used to check for this consistency. A consistent graph that is deadlock free has a *periodic schedule*. The period follows from the balance equations. By simulating the graph for at least a full period, the absence of deadlocks can be demonstrated.

The execution of the firing rule of node n takes an amount of time $t_n > 0$. Each firing has a start time and a finish time (= start time + t_n). A *timed schedule* is a function that returns the start time of the k-th firing of any node n. A timed schedule is *admissible* if, for every node in the graph, and for every firing of that node, the start time does not violate the firing rule. (A violation would be to schedule the consumption of a token before its production, specified by the finish time of the producing node.)

The *self-timed schedule* of a homogeneous SDF graph is the schedule where each node firing starts immediately once there is a token on each input edge. (And is therefore admissible.) A directed graph is said to be *strongly connected* if every node is reachable from every other node. A consistent and strongly connected SDF graph is deterministic, and its self-timed execution has the form of a lasso, an initial segment followed by a cycle (Ghamarian *et al.* 2006).

Let G be a homogeneous dataflow graph, containing a set of cycles G_C. The *cycle mean* of cycle c, denoted by $\mu(c)$, is defined by [8.2]

$$\mu(c) = \frac{\sum_{n \in N(c)} t_n}{\sum_{e \in E(c)} d_e}, \qquad [8.2]$$

where $N(c)$ and $E(c)$ denote the set of nodes and edges in cycle c, respectively; t_n denotes the execution time of node n; and d_e denotes the initial number of tokens on edge e. The *max cycle mean* of graph G is defined by [8.3]

$$\mu(G) = \max_{c \in G_C} \mu(c). \qquad [8.3]$$

The quantities $\mu(c)$ and $\mu(G)$ are also known as the *loop bound* of cycle c and the *iteration bound* of graph G. Reiter (1968) showed that once graph G is behaving periodically, the graph period equals $\mu(G)$ and the average throughput is $1/\mu(G)$. The reader is invited to calculate $\mu(G)$ for the three graphs in Figure 8.17 before reading on.

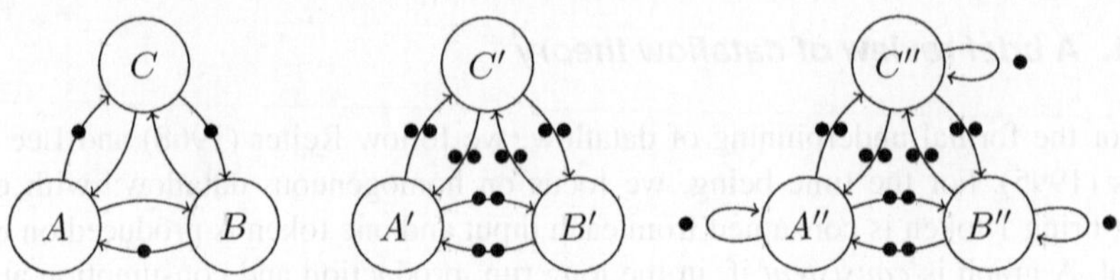

Figure 8.17. *Three tricycle dataflow graphs. Node execution times equal 1*

Figure 8.17 shows three similar dataflow graphs.[2] The leftmost graph has $\mu(G) = \infty$. The three cycles ABA, BCB, and CAC each have $\mu(c) = 2$.

2 The leftmost tricycle graph was introduced to me by Orlando Moreira.

However, there are two more cycles, namely $ACBA$ with $\mu(c) = 1$ and $ABCA$ without any initial tokens and hence $\mu(c) = \infty$. The graph deadlocks in its initial state as none of the nodes can fire. In the middle graph, all five cycles have $\mu(c) = \frac{1}{2}$, and hence $\mu(G) = \frac{1}{2}$. From its initial state onwards, the graph behaves periodically, the period equals $\frac{1}{2}$ and the firing rate of each node equals 2. Interestingly, the period is half the execution time of the individual nodes! Apparently, the dataflow formalism allows a node to start a next firing while it is busy with the previous one. This is also known as *auto concurrency*. Auto concurrency can be suppressed by introducing a self-loop for each node with one initial token. As a result, the rightmost tricycle graph has $\mu(G) = 1$. Adding back-edges and initializing self-loops are forms of dataflow graph augmentation.

The above theory can be extended to include static dataflow graphs, including cyclo-static graphs. Part of this extension relies on transforming graphs into SDF graphs or, further, into homogeneous dataflow graphs.

8.4.2. *The StaccatoLab execution model*

Building on the standard dataflow theory outlined above, we are now ready to describe the StaccatoLab execution model in terms of five choices.

First, the execution of the dataflow graph is *self-timed*, thus ensuring that all parallelism specified by the programmer is maintained. The direct implication is that there is no resource-sharing among different program nodes. Each node has its own resources to execute firing rules and to compute the values of the output tokens.

Second, node execution times are integer-valued, measured in *clock cycles*. A node with $t_n = 1$ takes one clock cycle to execute a firing rule. Auto concurrency can be avoided by adding a self-loop initialized with one token. When all nodes would have $t_n = 1$, this augmentation implies an alignment of all firing start and finish times with a global clock, which is intended to correspond to the hardware clock. However, $t_n > 1$ in combination with auto concurrency is essential to support non-trivial arithmetic functions (such as complex 32-bit multiplication) at acceptable clock frequencies. This form of pipelined execution is part of the execution model, but its formalization requires different (firing-) finish times per output.

Third, node firings can be *throttled*. Selected nodes can be equipped with a rate throttle. If the number of firings so far exceeds the number corresponding to a pre-set firing rate, otherwise enabled firings are postponed until the node is no longer ahead of its rate. This feature was used to real-time the token flow of the mp3 node in the audio example of Figure 8.8.

Fourth, node firings produce, at most, *one token* per firing per output. This simplifies both the implementation of nodes and the unfolding of (sub)graphs (section 8.2.3). Multi-rate nodes can consume an arbitrary number of tokens on each input but the production of multiple tokens takes multiple cycles. This does not limit node throughputs since a token value can be a tuple containing multiple values.

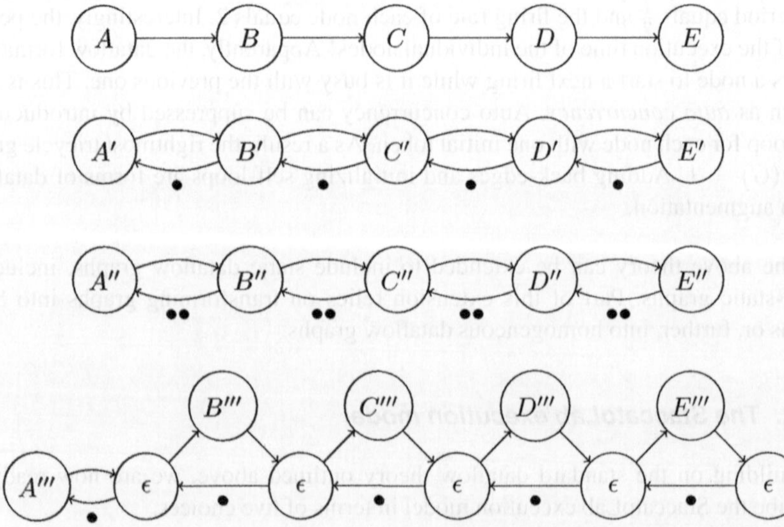

Figure 8.18. ***A dataflow pipeline with four different back-pressure schemes. Nodes A to E have an execution time of 1 and self-loops to bound the node's throughput to 1 (not drawn). ϵ-nodes have an execution time of $\epsilon \to 0$***

Fifth, and finally, we need to address *back-pressure*: how to bound the number of tokens on each edge. Dataflow as such does not bound the number of tokens. In the top graph of Figure 8.18, node A can keep firing indefinitely, even when node E is blocked from doing so, for whatever reason. The second pipeline graph is augmented by the introduction of a *back-edge* with one initial token, for each forward edge. (Note that with these back-edges, a connected graph becomes strongly connected.) In a homogeneous graph, the number of tokens in each cycle is invariant, and hence a forward edge can contain at most one token. As a result, $\mu(c) = 2$ for each cycle, limiting the throughput to $\frac{1}{2}$. This low throughput is overcome by initializing the back-edges with two tokens ($\mu(c) = 1$), at the expense of doubling the token-memory costs.

The StaccatoLab back-pressure scheme is based on the bottom graph in Figure 8.18. Graph augmentation introduces a so-called ϵ-node, with each node and the back-edge effectively bypassing the original node. As a result, $\mu(c) \leq 1 + 2\epsilon$, where ϵ is the execution time of the ϵ-node. If we assume $\epsilon \to 0$, this scheme has a throughput of 1, without the 2× penalty in token-memory costs. The four schemes are summarized in Table 8.2. The implementation of ϵ-nodes has its challenges, as outlined in section 8.5.2.

Back-pressure	$\mu(G)$	Throughput	Memory
No back-pressure	1	1	Unbounded
Back-edge with one initial token	2	$\frac{1}{2}$	$N-1$
Back-edge with two initial tokens	1	1	$2(N-1)$
ϵ back-edge	$1+2\epsilon$	$1/(1+2\epsilon)$	N

Table 8.2. *A comparison of the four back-pressure schemes of Figure 8.18*

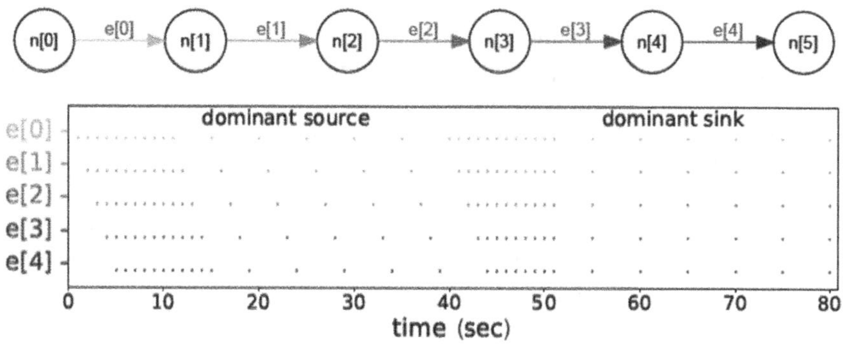

Figure 8.19. *A six-node pipeline starts with all nodes running self-timed with $t_n = 1$. After 11 firings, source node $n[0]$ is throttled to $t_{n[o]} = 5$. At time=40, the source throttle is switched off, and at time=50, the sink node is throttled to $t_{n[5]} = 5$*

We call this scheme *look-ahead back-pressure*. A node can fire when there is a vacancy on each output, which is a token on each back-edge. "Look-ahead" means that a node can also fire when these vacancies arise as a result of down-stream node firing *during the same clock cycle*. This has interesting implications. Forward tokens travel at a speed of one node per cycle. Vacancies (backward tokens) travel much lighter, with "the speed of light". See Figure 8.19. This means that a pipeline, after pausing due to back-pressure at its output, restarts fully and instantaneously during the first clock cycle after that back-pressure has been removed. Such back-pressure may arise, for example, when a pipeline is connected to a DRAM controller that also serves other clients. The StaccatoLab execution model is summarized in Table 8.3.

Feature	Brings
Self-Timed	Max parallelism and max throughput
Clocked	A simple performance model + a match with synchronous HW
Throttled	Real-timing of average throughput (throttle-up = unfolding)
One token per firing	A simple and efficient implementation ($\not\Leftrightarrow$ one value!)
Look-Ahead Back-pressured	Elastic pipelines with throughput=1 at lowest costs

Table 8.3. *A summary of the StaccatoLab execution model*

8.5. StaccatoLab

StaccatoLab is the name of the execution model outlined in section 8.4. It is also the name of a Python package that provides means to express, transform and analyze dataflow graphs, as well as to synthesize equivalent Verilog programs. The latter can then be synthesized for realization as ASIC or mapped onto one or more FPGAs.

8.5.1. *Dataflow graph description and analysis*

The StaccatoLab programming model is implemented as a Python API. The main Python classes visible to the programmer are `Edge`, `Node`, `FSM`, `Subgraph`, and `Graph`. Advantages of embedding the programming model in Python include the use of list comprehensions, polymorphism, parameterization and recursion, as well as a multitude of Python libraries to prepare and analyze a wide variety of data streams. Furthermore, a complete program, combined with various forms of analyses and multi-media descriptions, can be documented as a Jupyter Notebook (Jupyter 2020). A Jupyter notebook with all the programs described in this chapter can be found in van Berkel (2020).

The StaccatoLab simulator can be used in interactive mode or in batch mode. In the interactive mode, it supports cycle-by-cycle simulation, both forward and backward in time. In each state, any subgraph can be inspected at any detail, as well as the graphical representations of the node FSMs. The interactive simulator runs in a Jupyter cell. At any state, various views of the graph can be generated by means of Python methods called in other Jupyter cells. StaccatoLab also has a rich set of methods to prepare and analyze data streams, including audio streams and images.

For static dataflow graphs, the simulator detects a lasso by means of hashing the graph's state. Upon detection of a lasso, the simulator continues in the *lasso mode*, effectively replaying earlier state transitions. This leads to a simulation speed-up of 3-4x.

For the common case where data streams are partitioned in blocks, the internal structure of a subgraph can be replaced by a single node. A block can, for example, be an FFT input, an image or a finite list of key-value pairs. In the *block mode*, the output function `fo` of such a node is evaluated on the block as a whole, but tokens are consumed and produced individually. This enables mixed-level simulation, where either an existing subgraph is replaced by a node to speed-up simulation, or a node acts as an abstraction of a subgraph yet to be designed.

8.5.2. *Verilog synthesis*

Verilog is a hardware description language commonly used in the design and verification of digital circuits at the register-transfer level of abstraction. This section describes the mapping of dataflow graphs onto Verilog programs, assuming the execution model of section 8.4. At this stage, it is too early to report performance numbers and meaningful benchmarking, as the current Verilog mapper lacks several features and optimizations.

The self-timed execution of a dataflow graph implies that each node has its own, dedicated hardware resources. The synthesis of Verilog is modular, organized node by node, where the input edges to a node are included in the node's Verilog module. The structure of such a Verilog module is depicted in Figure 8.20.

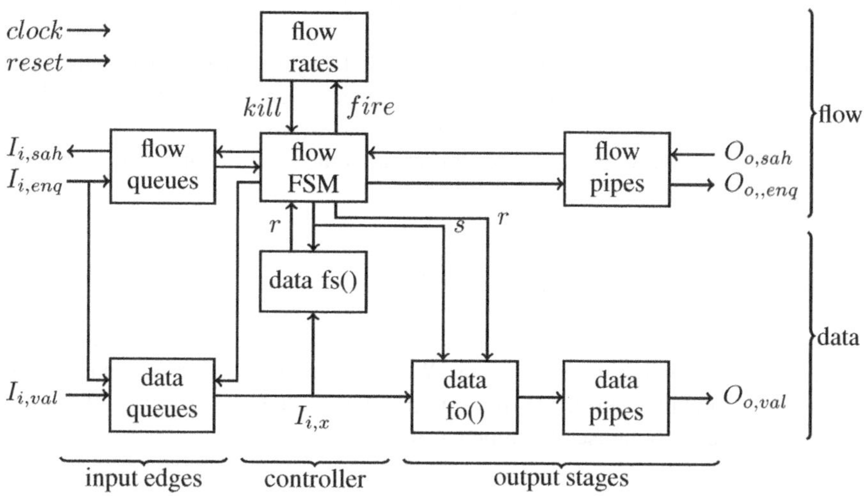

Figure 8.20. *Verilog template for a StaccatoLab node,* including input edges

Submodule *flow queues* keep track of the number of tokens and slack for each input queue. During each clock cycle, signal I_{sah} specifies the amount of *slack ahead* (the number of back tokens) and I_{enq} specifies the number of tokens to be added to the queue. The token values are stored in submodule *data queues*. The data queue may range from a single value or a few values (stored in Verilog registers), to a large array with last-produced and last-consumed pointers that are incremented based on the number of tokens produced and consumed for each input edge.

The FSM submodule can range from a few gates for a homogeneous dataflow node to an elaborate state machine with multiple counters to support CSDF, a rule selector for data-dependent dataflow, or possibly a round-robin scheduler to support non-determinate dataflow. To check whether or not a firing rule is enabled, the state of node input queues and the values of O_{sah} are inspected, and when firing takes place, the appropriate O_{enq} values are produced. Optional submodule *flow rates* count firings relative to clock cycles, and kill an otherwise enabled firing rule when the node runs ahead of its set rate.

Submodule *data fs()* implements the `Select` function. It is a combinational function that has access to the input data, and its output determines the next `Select` firing rule, provided it is enabled by the required input tokens.

Submodule *data fo()* is typically an arithmetic function, computing one or more values from the data queues as well as the FSM state. The depth of the output pipeline determines the execution time. Optional submodule *flow pipes* and *data pipes* define pipelines to support pipelined execution of *data fo()*.

The look-ahead back-pressure scheme is challenging because it is intrinsically non-local. The back propagation of the slack-ahead signal is a combinational path from O_{sah} to I_{sah}. Special measures are required to avoid this path becoming the critical path. A variety of techniques are used, some akin to carry acceleration, as applied to adder circuits.

8.6. Large-scale dataflow computing?

In this concluding section, we make the case for large-scale dataflow computing. We do so by addressing three questions: What kind of applications is dataflow good for? Why is dataflow effective as a programming model? Why would dataflow programs make efficient use of compute resources?

8.6.1. *What kind of applications?*

An application is *embarrassingly parallel* when parallelism is abundant *and* there are no or few dependencies between the subtasks, often called *threads*. An example of an embarrassingly parallel application is ray tracing (3D computer graphics): each pixel can be computed in parallel, independent of all other pixels. The intra-stage and multi-FFT parallelism discussed in section 8.1, as well as matrix–matrix multiplication, are other examples of embarrassingly parallel tasks. All parallel programming models can deal with this form of parallelism.

Dependencies among threads require synchronization. In GPUs, such synchronization is applied at a coarse granularity; for example, when *all* threads of FFT stage i of FFT j are completed, the program can proceed to stage $i + 1$ of FFT j. Dataflow is very effective in capturing such synchronization at any granularity, for example, between two tokens ready to be "joined" by an individual FFT butterfly. It is this fine-grained synchronization that led to the early application of dataflow in the domain of signal processing, including audio, video and telecommunication (both wired and wireless). Probably, the best known example here is LabView (Andrade and

Kovner 1998). LabView supports the description of signal processing applications in a *graphical* dataflow language and their synthesis as embedded software code or as FPGA hardware. LabView covers (cyclo-)static dataflow, as well as limited forms of dynamic dataflow, not unlike Gao *et al.* (1992).

Maxeler Technologies applies dataflow well beyond signal processing, even aspiring to "exascale performance". For Becker *et al.* (2017): "Porting an application onto a dataflow system requires the application code to be split into a host application and a dataflow part." The "control-oriented" host application is mapped onto CPU(s) and the dataflow part is mapped onto so-called dataflow engines (DFEs). A DFE is a rather coarse-grained, data-independent, "laid-out in space", "deeply pipelined compute structure". Their impressive *application gallery* (Maxeler Technologies 2019) lists applications in data analytics, engineering, transaction processing, networking, science and security.

My personal ambition is to apply dataflow programming techniques to radio astronomy applications, including imaging (van Berkel 2015, 2016) and pulsar search (van Berkel and Boerman 2019).

8.6.2. *Why effective?*

The effective design of parallel programs requires not only means to express multiple forms of parallelism, but also means to *quantitatively reason* about the performance of these programs. Performance (throughput) in today's HPC is a quantity that you *measure* when your program runs on a given machine. This performance is the result of the execution of compiled multi-threaded code, dynamic thread-block scheduling, barrier synchronization and other forms of synchronization of code running on that machine. In the event you are not satisfied with the measured throughput, you can modify parts of the code, rerun and remeasure. This heterogeneity of scheduling and synchronization mechanisms makes it very difficult to quantitatively analyze the throughput based on the program.

Dataflow programming, in contrast, supports quantitative reasoning about program throughput. The FFT bit-reversal program of section 8.2.2 has a throughput of one token per clock cycle, and can be transformed into a program that accepts, for example, four tokens each clock cycle. This performance analysis is ultimately based on the dataflow theory outlined in section 8.4.1. For this theory, it does not matter whether a token carries a boolean, a complex number, a vector, a matrix or an image. As such, dataflow formalism scales nicely over orders of magnitude in the size of compute tasks. Moreover, dataflow not only supports the quantitative analysis of resource needs, but it also enables a formal and structured trade-off among resources. For example, the (on-chip) memory costs of the recursive merge sorter of section 8.3.1 can be reduced by up to 50% at the expense of a proportional decrease in throughput.

Together with (partially) automated graph transformations (section 8.2.3), this invites the design of programs for a given target rate.

8.6.3. *Why efficient?*

Compute is *not* the scarce resource in HPC. After all, 97% of the arithmetic resources is routinely idling, as shown in section 8.1. In order to involve a larger fraction of these arithmetic resources, the challenge is to move the operands to the right place at the right time. The scarce resources for HPC are therefore the amount on-chip memory and the bandwidths to both on-chip memory and off-chip memory. GPUs spend the scarce on-chip memory on huge register files and caches, in ways that offer little transparency to the programmer, and that allow little control by the programmer. As a result, the on-chip memory is used ineffectively and the off-chip memory traffic may increase by as much as 5× (section 8.1). Dataflow by its very essence offers a dedicated queue for each operand for each operator, and by virtue of the execution model of section 8.4, these queues are managed at a clock-cycle granularity.

A dataflow node implemented as in Figure 8.20 can be viewed as an application-specific core. Such a core can be small (e.g. a counter) or fairly large (e.g. a 4x-SIMD complex-FFT butterfly). Likewise, the controlling FSM can be small (for homogeneous dataflow) or elaborate (for CSDF and dynamic dataflow). These *customized* cores fit nicely with the heterogeneous resource mix of FPGAs. Similarly, the multiple registers in the FPGA DSP blocks, together with the block RAMs, nicely support the variety in buffer sizes and data types. Very large edges can be (partially) mapped onto off-chip DRAM, and large graphs can be partitioned over multiple FPGAs using inter-FPGA interconnect. On the downside, FPGAs are typically clocked 2–3× slower than GPUs. But again, as the bottleneck is often in the off-chip bandwidth, this may not matter that much. Also, Lant *et al.* (2019) make the case for FPGA-based HPC, hinting at "data-flow type processing".

Moving 64b from an off-chip DRAM to an on-chip adder costs 1 nJ in energy. Moving the same item from a 20 MB on-chip memory costs 100 pJ, from an 8 kB memory 10 pJ, and from a local register less than 1 pJ. These numbers apply to 45 nm CMOS and are from Figure 1.1.9 in Horowitz (2014). Ultimately, this is probably the most compelling reason for why a programmer may want to control the flow of data. For some applications, a 10× power reduction is to be gained (van Berkel 2015).

The proposed dataflow programming and execution models offer the programmer the means to explicitly control the flows of data across this memory hierarchy. Moreover, they encourage the exploration of novel algorithms to deal with the scarce resources of today's computing technology, both effectively and efficiently.

8.7. Acknowledgments

I am indebted to Rudolf Mak for his continuous and constructive feedback during the development of the ideas presented. This chapter also benefited from his review.

8.8. References

Andrade, H.A. and Kovner, S. (1998). Software synthesis from dataflow models for G and LabVIEW/sup TM/. *Conference Record of 32nd Asilomar Conference on Signals, Systems and Computers (Cat. No.98CH36284)*, 2, 1705–1709.

Arvind and Brock, J.D. (1984). Resource managers in functional programming. *Journal of Parallel and Distributed Computing*, 1(1), 5–21 [Online]. Available: http://www.sciencedirect.com/science/article/pii/074373158490008X.

Becker, T., Burovskiy, P., Nestorov, A.M., Palikareva, H., Reggiani, E., and Gaydadjiev, G. (2017). From exaflop to exaflow. *Design, Automation Test in Europe Conference Exhibition*, pp. 404–409.

van Berkel, K. (2015). Astronomical workloads [Online]. Available: https://mpsoc-forum.org/archive/2015/.

van Berkel, K. (2016). Exascale computing for radio astronomy: GPU or FPGA [Online]. Available: https://mpsoc-forum.org/archive/2016/.

van Berkel, K. (2018). Mapping dataflow graphs to circuits [Online]. Available: https://mpsoc-forum.org/archive/2018/.

van Berkel, K. (2020). StaccatoLab: A programming and execution model for large-scale dataflow computing [Online]. Available: https://www.win.tue.nl/~cberkel/StaccatoLab-program.html.

van Berkel, K. and Boerman, F. (2019). Exascale computing for radio astronomy: Mapping pulsar search on dataflow [Online]. Available: https://mpsoc-forum.org/archive/2019/.

Bilsen, G., Engels, M., Lauwereins, R., and Peperstraete, J.A. (1995). Cyclo-static data flow. *1995 International Conference on Acoustics, Speech, and Signal Processing*, 5, 3255–3258.

Buck, J.T. and Lee, E.A. (1993). Scheduling dynamic dataflow graphs with bounded memory using the token flow model. *IEEE International Conference on Acoustics, Speech, and Signal Processing*, 1, pp. 429–432.

Dennis, J.B. (1974). First version of a data flow procedure language. In *Programming Symposium*, Robinet, B. (ed.), Springer, Berlin, Heidelberg.

Gao, G.R., Govindarajan, R., and Panangaden, P. (1992). Well-behaved dataflow programs for DSP computation. *IEEE International Conference on Acoustics, Speech, and Signal Processing*, 5, 561–564.

Garrido, M., Grajal, J., and Gustafsson, O. (2011). Optimum circuits for bit reversal. *IEEE Transactions on Circuits and Systems II: Express Briefs*, 58(10), 657–661.

Garrido, M., Grajal, J., Sanchez, M.A., and Gustafsson, O. (2013). Pipelined radix-2^k feedforward FFT architectures. *IEEE Transactions on Very Large Scale Integration (VLSI) Systems*, 21(1), 23–32.

Garrido, M., Acevedo, M., Ehliar, A., and Gustafsson, O. (2014). Challenging the limits of FFT performance on FPGAS. *2014 International Symposium on Integrated Circuits (ISIC)*, pp. 172–175.

Ghamarian, A.H., Geilen, M.C.W., Stuijk, S., Basten, T., Theelen, B.D., Mousavi, M.R., Moonen, A.J.M., and Bekooij, M.J.G. (2006). Throughput analysis of synchronous data flow graphs. *Sixth International Conference on Application of Concurrency to System Design*, pp. 25–36.

Govindaraju, N.K., Lloyd, B., Dotsenko, Y., Smith, B., and Manferdelli, J. (2008). High performance discrete fourier transforms on graphics processors. *Proceedings of the 2008 ACM/IEEE Conference on Supercomputing*, pp. 1–12.

Horowitz, M. (2014). 1.1 computing's energy problem (and what we can do about it). *2014 IEEE International Solid-State Circuits Conference Digest of Technical Papers (ISSCC)*, pp. 10–14.

HPCG (2019). HPCG benchmark [Online]. Available: http://www.hpcg-benchmark.org.

Johnston, W.M., Hanna, J. R.P., and Millar, R.J. (2004). Advances in dataflow programming languages. ACM Computing Surveys, 36(1), 1–34 [Online]. Available: http://doi.acm.org/10.1145/1013208.1013209

Jupyter (2020). The Jupyter Notebook [Online]. Available: https://jupyter.org.

Karp, R.M. and Miller, R.E. (1966). Properties of a model for parallel computations: Determinacy, termination, queueing. *SIAM Journal on Applied Mathematics*, 14(6), 1390–1411 [Online]. Available: https://doi.org/10.1137/0114108.

Lant, J., Navaridas Palma, J., Lujan, M., and Goodacre, J. (2019). Making the case for FPGA based HPC. *IEEE Micro*, 1–1.

Lee, E.A. (1991). Consistency in dataflow graphs. *IEEE Transactions on Parallel and Distributed Systems*, 2(2), 223–235.

Lee, E.A. (2006). The problem with threads. *Computer*, 39(5), 33–42.

Lee, E.A. and Messerschmitt, D.G. (1987). Synchronous data flow. *Proceedings of the IEEE*, 75(9), 1235–1245.

Lee, E.A. and Parks, T.M. (1995). Dataflow process networks. *Proceedings of the IEEE*, 83(5), 773–801.

Maxeler Technologies (2019). Dataflow applications [Online]. Available: http://appgallery.maxeler.com.

Parhi, K. (2007). *VLSI Digital Signal Processing Systems: Design and Implementation*. Wiley India Pvt. Limited [Online]. Available: https://books.google.nl/books?id=APFRHFkMqG8C.

Parhi, K.K. and Messerschmitt, D.G. (1991). Static rate-optimal scheduling of iterative data-flow programs via optimum unfolding. *IEEE Transactions on Computers*, 40(2), 178–195.

Ptolemaeus, C., (ed.) (2014). *System Design, Modeling, and Simulation using Ptolemy II*, Ptolemy.org [Online]. Available: http://ptolemy.org/books/Systems.

Reiter, R. (1968). Scheduling parallel computations. *Journal of the ACM*, 15(4), 590–599 [Online]. Available: http://doi.acm.org/10.1145/321479.321485.

TOP500 (2019). Top500, the list [Online]. Available: https://www.top500.org.

Williams, S., Waterman, A., and Patterson, D. (2009). Roofline: An insightful visual performance model for multicore architectures. *Communications of the ACM*, 52(4), 65–76. [Online]. Available: http://doi.acm.org/10.1145/1498765.1498785.

Wong, H., Papadopoulou, M., Sadooghi-Alvandi, M., and Moshovos, A. (2010). Demystifying GPU microarchitecture through microbenchmarking. *2010 IEEE Int. Symp. on Performance Analysis of Systems Software (ISPASS)*, pp. 235–246.

9

Smart Cameras and MPSoCs

Marilyn WOLF

Department of Computer Science and Engineering, University of Nebraska–Lincoln, USA

Embedded computer vision is an important application area for multi-processor systems-on-chips (MPSoCs). Smart cameras make use of MPSoCs to perform real-time computer vision, requiring high-performance computing and memory. The capabilities and architectures of MPSoCs and smart cameras have advanced in parallel. This chapter examines the development of MPSoCs for video and computer vision. As computer vision algorithms have developed, architectures have developed to support the efficient execution of computational kernels and to couple multiple parallel units to implement complex vision pipelines.

9.1. Introduction

Embedded computer vision performs complex computer vision operations in real time. Not all computer vision applications require real-time performance – image classification for web applications, for example, may require high performance but not real-time behavior. Embedded computer vision places complex demands on a computing platform: high computation rates, high memory bandwidth and complex data flows. Building a real-time vision system requires careful design and matching of application requirements, algorithms and computing platforms.

Computer vision tasks include subject identification (*is there something in front of the camera?*), classification (*what is in front of the camera?*) and tracking (*where*

For a color version of all figures in this book, see www.iste.co.uk/andrade/multi2.zip.

Multi-Processor System-on-Chip 2 – Applications,
coordinated by Liliana ANDRADE and Frédéric ROUSSEAU.

is the subject moving?). Figure 9.1 shows a pair of visible and infrared images from the CAMEL dataset (*https://camel.ece.gatech.edu/*) along with the results of classification. The labels provide both a type and a confidence for the classification.

visible *infrared*

Figure 9.1. *Visible and infrared images and associated classification from the CAMEL dataset (*https://camel.ece.gatech.edu*)*

This chapter examines the development of multi-processor system-on-chip technology from the viewpoint of real-time computer vision. The high computing and memory demands of computer vision make it a good subject for MPSoC design. This chapter is not intended as a comprehensive study of all processors that could be used for vision. A few designs are used to trace the requirements that drive design and the architectural approaches deployed to solve those problems.

The next section describes early video processors designed for video encoding. Section 9.3 examines programmable video processors. Section 9.4 examines accelerators for subsystems of computer vision systems. Section 9.5 considers the evolution of DSP and VLIW approaches to heterogeneous multi-processor systems-on-chips. Section 9.6 describes graphic processing units (GPUs) as vision processors. Section 9.7 shows how neural networks motivated the introduction of tensor processing units. Section 9.8 presents some concluding remarks.

9.2. Early VLSI video processors

Video compression was the driving application for an early generation of VLSI video processors. A variety of approaches were used: pipelining, SIMD or array processors and multi-processors connected by point-to-point and/or bus connections.

Nomura *et al.* (1989) designed a parallel DSP for video encoding. Their compression algorithm was based on discrete cosine transform analysis and coefficient quantization, similar to methods used in modern JPEG. Their system included 72 processing elements, each based on a T800 transputer with 10 MIPS of performance and 1 megabyte of memory.

Inoue *et al.* (1993) designed a video signal processor based on BiCMOS technology. The processor operated at 300 MHz. The processor could encode CIF (352×288-pixel) video at 30 frames/s using H.261. The processor included seven address generation units, a 16-bit arithmetic logic unit and a pipelined convolver unit. The processor was heavily pipelined to perform vector operations.

Ackland *et al.* (1993) designed an early video codec chip set; they also identified speech recognition and synthesis and handwriting recognition as possible applications. Their complete codec included three custom chips and one or two DSPs. The DSPs were used for audio coding/decoding, JPEG still picture decoding and echo cancelation. A systems controller chip manages external memory and devices. Two chips, one for encoding and the other for decoding, provide core video compression/decompression functions based on the $P \times 64$ (up to 288×360 pixels) and MPEG (up to $1{,}024 \times 1{,}024$ pixels) standards. The decoder includes a SIMD signal processor as well as special-purpose function units for operations such as variable-length decoding. The encoder includes a motion estimator a SIMD signal processor, operators for variable length encoding, as well as a memory controller and output FIFO.

Araki *et al.* (1994) developed a video DSP for MPEG-2 encoding and decoding. Their processor used SIMD to process 8×8 blocks, a pipeline organized around the MPEG block diagram for the overall video stream. A pair of chips could encode MPEG-2, and a single chip could decode MPEG-2.

Miyake *et al.* (1998) developed a systolic array for high-definition video decoding. The architecture was designed to support 1D and 2D filtering, pixel differencing and min/max detection. The video processing elements could be programmed but were not Turing-complete – they lacked branch or memory access instructions.

9.3. Video signal processors

The video signal processor (VSP) emerged as a way to provide performance and high programmability to video processing. A VSP is a digital signal processor optimized for video workloads. The term was originally applied primarily to single-issue DSPs or vector processors optimized for video and came to be used for VLIW video processors.

The Texas Instruments MVP (Instruments 1995) was an early commercial VSP. As shown in Figure 9.2, the VSP was a crossbar-connected multi-processor with four parallel processors (PPs) and an RISC master processor. Each parallel processor was a long instruction word unit that could dispatch a data operation, load and store from a single instruction. The data ALU was augmented with operators for specialized image and video operations. The RISC CPU included a floating-point unit. A transfer controller interfaces to external memory.

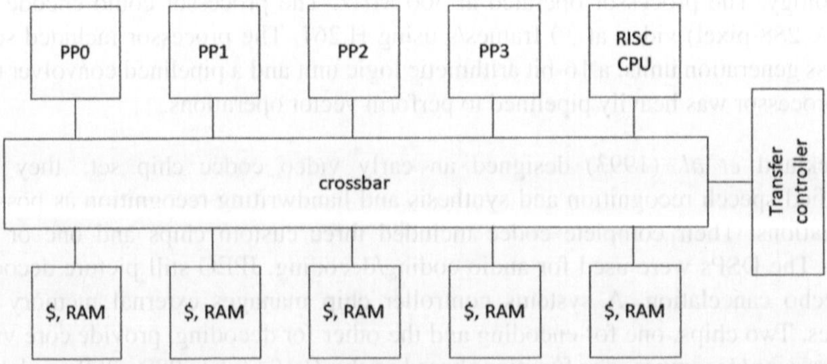

Figure 9.2. *The TI MVP*

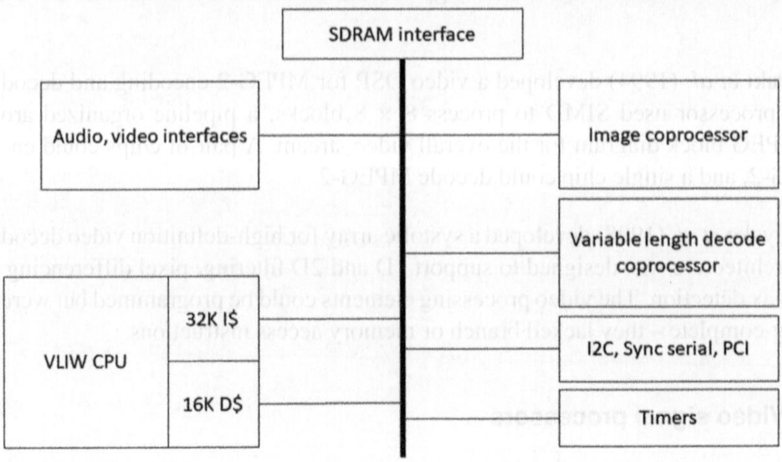

Figure 9.3. *The Trimedia TM-1*

The Trimedia TM-1 (Slavenburg *et al.* 1996) appeared slightly after the MVP. As shown in Figure 9.3, the TM-1 is a typical processor-plus-bus configuration. A memory interface connects to external SDRAM. Interfaces include audio, video, PCI, synchronous serial and I^2C. Co-processors for variable-length decoding, up/downscaling and color space conversion are included. The VLIW processor is a five-issue unit with a 128-word register file and 27 function units.

Jeschke *et al.* (Pirsch and Gehrke 1995) developed a methodology to model performance of multi-processors for video processing; the multi-processor model included SIMD and MIMD processing elements. The methodology allowed for the optimization of silicon area for a given video source rate. Dutta *et al.* (1998) analyzed a family of VLIW video signal processors designed to support programming in high-level languages. They evaluated a design space in the range of 32–64 operations per cycle. They concluded that crossbars were practical up to 32 ports. They also concluded that local memory introduced fewer bottlenecks than both highly multiported memories and banked global memories.

9.4. Accelerators

Accelerators perform well-defined operations, perhaps with some parameterization. Accelerators generally provide some combination of performance and power grains, assuming that they can be effectively deployed within a multi-processor. Computer vision defines a number of computationally intensive operations that are well suited to implementation as accelerators.

Block motion estimation (Netravali and Robbins 1979) is a key operation in video compression. Motion estimation compares a current frame $c(I_c, J_c)$ with a previous frame $p(I_p, J_p)$ given an offset $\langle o_x, o_y \rangle$, as shown in [9.1]

$$S(o_x, o_y) = \sum_i \sum_j |c(I_c + i, J_c + j) - p(I_p + i + o_x, J_p + j + o_y)| \quad [9.1]$$

Motion estimation makes use of correlation, an operation that has uses in both algorithmic and neural network computer vision.

Early motion estimation systems used full search over a given window. Full search requires a great deal of memory bandwidth. As a result, a wide range of motion estimation search algorithms have been developed using different search patterns that test some but not all of the positions in the search window. Motion estimation is amenable to SIMD operation. Yang *et al.* (1989) developed a parallel architecture for motion estimation that evaluates one row of the motion estimation at a time. A simplified version of their architecture is shown in Figure 9.4. The pixels from the current frame are shifted vertically to test different positions, while the previous frame pixels are held at a fixed processing element.

Yang *et al.* (2005) analyzed the power and performance of several realizations of motion estimation algorithms. They evaluated both hardware and software implementations. They found that several search algorithms resulted in similar power consumption levels, all considerably below full search. They also found that the subsampled motion field with the alternating pixel-decimation search algorithm (Liu and Zaccarin 1993) was the most efficient when realized in software but similar to the one-dimensional full search algorithm (Chen *et al.* 1994) when implemented in hardware.

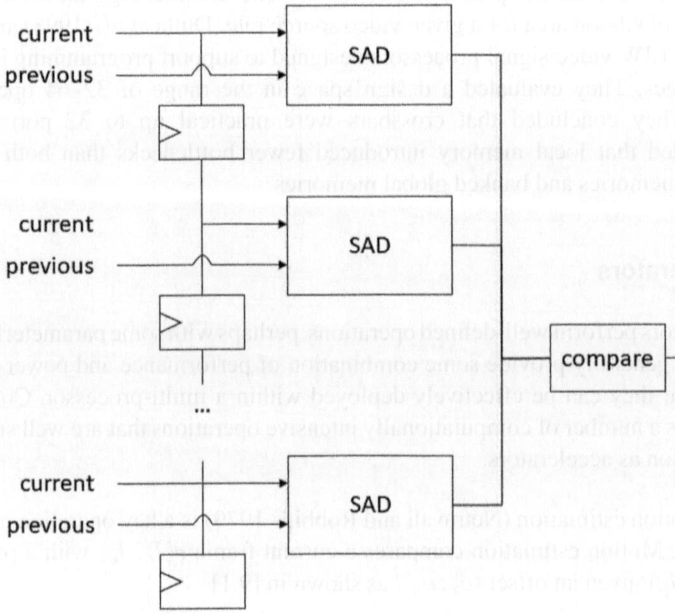

Figure 9.4. *Motion compensation accelerator*

Schlessman *et al.* (2005) developed an accelerator for shape detection inspired by Yang's motion estimation architecture. Multiple processing elements perform several correlations in parallel. After the search area has been fully evaluated, the best match is selected. The parallelism of the search can be adjusted using a sequencer and address generator. The time required to complete the search depends on the number of processing elements used.

Chellappa *et al.* (2005) developed an accelerator for face detection. The detector assumes an elliptical model for the face. Hardware synthesis from a signal flow graph was used to explore the design space for the accelerator.

Schlessman *et al.* (2007) developed accelerators for optical flow using the Kanade–Lucas–Tomasi (KLT) algorithm (Lucas and Kanade 1981; Tomasi and Kanade 1991) and for background subtraction using mixture-of-Gaussian models (Horprasert *et al.* 1999). The Kanade–Lucas–Tomasi algorithm uses convolution to determine partial derivatives I_x, I_y, I_t for the difference between current and previous frames. It then computes the motion vector according to [9.2]

$$V = \begin{bmatrix} V_x \\ V_y \end{bmatrix} = \left[A^T A\right]\left[-A^T I_t\right] \qquad [9.2]$$

Mixture-of-Gaussian maintains a fixed number of pixel-by-pixel Gaussian models. Each model includes mean and variance for the three color channels, such as luminance Y, blue chrominance C_b and red chrominance C_r. It selects the model that best fits background characteristics based on the variation of brightness distortion, as shown in [9.3]

$$\alpha_x = \sqrt{\frac{|X - \mu_X|}{\sigma_X}} \qquad [9.3]$$

where $x \in \{Y, C_r, C_b\}$. These three values are used to test whether a pixel is considered foreground or background. The algorithm also updates the models. Schlessman and Wolf (2015) analyzed the numerical precision required for these optical flow and mixture-of-Gaussian background elimination.

Feature classification can be used as an intermediate phase in a number of different computer vision algorithms: object correspondence, classification, etc. The SIFT algorithm (Lowe 1999) is widely used to generate robust features – it provides a scale-invariant feature transform. SIFT operates in several steps: Gaussian smoothing is used to generate an image pyramid at several different scales; key points are identified at local maxima or minima; histogram analysis is used to identify the orientation of the feature; a descriptor is generated that describes the feature in terms of several subregions. Huang *et al.* (2012) developed a hardware architecture for SIFT with blocks corresponding to the major stages of the algorithm.

9.5. From VSP to MPSoC

Video signal processors were modeled on traditional multi-processors and micro-processors. Multi-processors were fairly symmetric; VLIW video processors leveraged the ease with which video operations could be statically scheduled. Both performance and power encouraged designers to consider more heterogeneous approaches with accelerators complementing programmable processors.

The Philips Viper (Dutta *et al.* 2001) used the Trimedia as part of a multi-processor architecture targeting MPEG-2; it is also most likely the first commercially available MPSoC (Wolf *et al.* 2008). As shown in Figure 9.5, Viper combines a Trimedia processor and MIPS CPU interconnected with four busses. A variety of accelerators are available. A memory controller connects to external SDRAM.

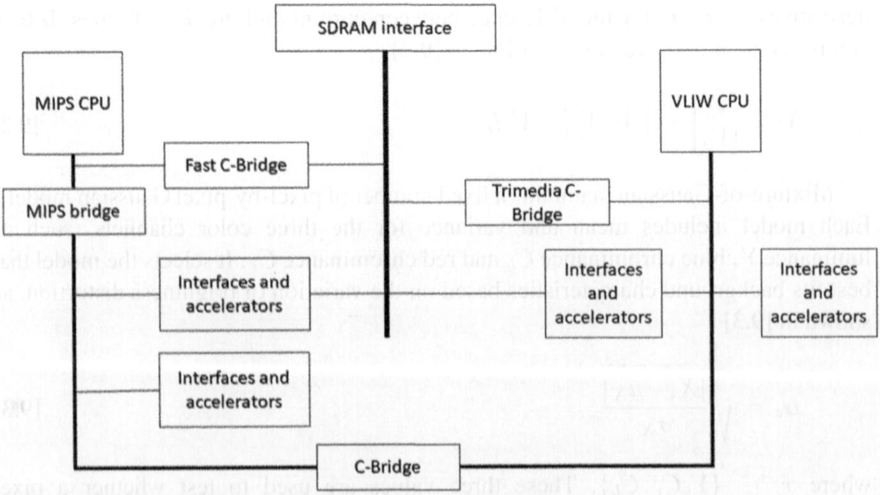

Figure 9.5. *The Philips Viper*

The Texas Instruments DaVinci (Instruments 2010), introduced in 2005, is an example of a later approach to a VSP system-on-chip. The VLIW processor includes six ALUs and four multipliers. An ARM9 processor executes a general-purpose code. A video processing subsystem interfaces to imagers, performs color space conversion, histogram generation, image resizing and other functions. A video-imaging co-processor provides a DSP-based unit for additional video operations. The processing elements are connected by a switched central resource.

Given the high data rates required for computer vision, interconnect is a key concern; quality-of-service requirements put further emphasis on the interconnect. Xu *et al.* (2004) analyzed various interconnect structures for an architecture to implement the smart camera algorithms of Wolf *et al.* (2002). The vision system was mapped onto five processors: P0, P2 and P4 form one pipeline; P1, P3 and P5 form another pipeline; and both pipelines feed the final P6 stage. Their study compared a bus and several versions of crossbar interconnect. One set of crossbars were sized as 10×10 or 11×11 for single and dual memory, respectively; the crossbars varied in their width, ranging from 5 bits per port to 55 bits per port. They simulated the networks using OPNET and traces from the smart camera system. They also evaluated a 3×3 crossbar with an input queue. They found that a 3×3 crossbar with two shared memories allowed the application to run at 150 frames/s; this configuration provided the lowest maximum throughput, which is associated with a small area.

Sen *et al.* (2005) synthesized the smart camera algorithms of Wolf *et al.* from a homogeneous parameterized dataflow graph. They synthesized the design onto a Xilinx Virtex II FPGA.

Face detection is widely used in applications ranging from surveillance to camera autofocus. Theocharides *et al.* (2004) developed an architecture for embedded face detection. The system included several stages:

– a filter unit created an image pyramid with resolutions ranging from 300 × 300 to 20 × 20;

– a histogram equalization unit adjusted the exposure of each 20 × 20 window;

– a rotation detection network identified the rotation of a possible face;

– an image rotation unit performs the required rotation for normalization;

– a brightness and contrast adjustment unit adjusts contrast to enhance facial characteristics;

– a set of three detection neural networks performs the final face detection. Each unit contains three parallel stages, each designed to identify a different set of component features, along with a final classification stage.

9.6. Graphics processing units

Graphics processing units have been widely used for traditional algorithmic computer vision systems; they have also been used for artificial neural networks, as described below. Graphics processing units provide streaming multi-processors and large amounts of internal memory bandwidth that can be leveraged by data-parallel applications such as computer vision. GPUs also traditionally provide several different levels of floating-point precision, allowing designers to trade off precision, performance and power consumption.

Mobile GPUs leverage desktop GPU architectures with smaller numbers of processing elements for low-power mobile applications. The NVIDIA Jetson TX2 (*https://www.nvidia.com/en-us/autonomous-machines/embedded-systems/jetson-tx2/*) includes a Pascal-architecture GPU with 256 CUDA cores. Its CPU complex includes a dual-core Denver 2 64-bit CPU and a quad-core ARM A57. The NVIDIA AGX Xavier includes a Volta-architecture GPU with 384 CUDA cores and 48 tensor cores. Its CPU complex includes a six-core, 64-bit Carmel ARM. It also includes a seven-way VLIW vision processor.

9.7. Neural networks and tensor processing units

Neural networks have emerged as a very different approach to computer vision than that of the traditional algorithmic approaches. Artificial neural networks have been studied for decades. They have received a great deal of recent attention, as evidenced by the ACM Turing Award given to Benigno, Hinton and LeCun in 2018 (ACM 2018). Many formulations of neural networks combine convolution with a nonlinear operation. AlexNet is a neural network designed for implementation on a pair of GPUs (Krizhevsky *et al.* 2012); most of the layers are split between the two GPUs. This network demonstrated that large training sets could be evaluated by GPUs. The size of correlation windows and the number of distinct windows vary by layer.

Memory system performance is a key design issue for digital neural networks given the large numbers of coefficients required by modern networks. Neural networks are used for inference by providing input values and computing their resulting output. The coefficients in the network are derived by training. These two operations present very different memory access patterns. Inference operations in a network such as AlexNet can be fully statically scheduled and arranged to maximize memory system parallelism. Inference computations move forward through the network from inputs to outputs. Training involves both forward and backward data dependencies. When an output is identified as requiring modification, the training algorithm traces backward to identify weights that should be modified to change the output. These operations are not amenable to static scheduling.

Training is often performed using floating-point values. Inference, in contrast, can in many cases be performed using integer arithmetic for convolution and approximations for nonlinear functions. Zhuang *et al.* (Wren *et al.* 1997) developed a methodology to optimize the required bit width. The FINN-R system (Blott *et al.* 2018) optimizes neural networks for implementation on FPGAs by evaluating opportunities for quantization. Prost-Boucle *et al.* (2018) developed an inference architecture based on ternary $\{-1, 0, 1\}$ weights.

Chen *et al.* (2016) developed an accelerator for large neural networks. Their processor provided buffers for inputs, outputs and synaptic weights. A neural function unit provided a set of datapaths for parallel convolutions and nonlinear operations.

Recent efforts to execute neural networks at high rates have concentrated on the convolution operation as formulated as tensor operations. A tensor describes a linear relationship between algebraic objects. A common tensor formulation for a neural network makes use of a 4×4 matrix multiply/accumulate, as shown in [9.4].

$$R = AW + C \qquad [9.4]$$

Memory systems are designed to feed the tensor units based on the structure of the neural network being evaluated.

Tensor processing units have been designed for both data centers, and mobile operations. Data centers must be able to perform both training and inference.

The NVIDIA GV100 (Tesla 2017) provides an example of a modern processor with tensor unit accelerators. The V100 tensor cores perform mixed-precision arithmetic: A and W are 16-bit floating-point values (FP16), while C and R are

32-bit floating point values (FP32). A tensor core performs 64 fused multiply-accumulates per clock cycle. The V100 includes 80 streaming multi-processors. Each streaming multi-processor includes eight tensor cores for a total of 640 for the entire GPU. The Baidu neural processor (Shilov 2019) couples processing cores on one chip with a pair of HBM2 stacked memory packages. Cerebras developed a wafer-scale neural processor (Systems 2019): 46,225 mm^2 of silicon with 1.2 trillion transistors in a 16 nm process. The chip is designed to support training operations that require higher levels of numerical precision and more flexible memory accesses. A set of 400,000 fully programmable processing elements includes both general operations and machine learning extensions, including tensor optimizations. Moreover, 18 GBytes of memory is distributed on-chip. Processing elements and memory are connected in a 2D mesh.

Mobile units are often used primarily for inference. The Apple A13 (Frumusanu 2019a) and Samsung Exynos 990 (Frumusanu 2019b) are examples of smartphone processors with neural processing units.

9.8. Conclusion

MPSoCs for computer vision have seen several rounds of evolution. Early video processors, designed primarily for video coding, explored both heavily pipelined units and SIMD operation. The need for programmability resulted in the development of video signal processors based on VLIW architectures. As computer vision emerged as an important application for MPSoCs, a variety of accelerators were developed for computer vision kernels. These kernels were incorporated into heterogeneous MPSoC architectures. Graphics processing units provided programming flexibility as well as high performance. The emergency of neural networks for computer vision has, for the moment, resulted in architectures that emphasize linear algebra operations over huge amounts of data. These tensor processors have been developed at huge scales for data center use and at smaller scales for portable operation. We should expect these architectures to evolve further as applications push more training to the edge of the network.

9.9. References

Ackland, B.D., Aghevli, R., Eldumiati, I., Englander, A.C., and Scuteri, E. (1993). A video-codec chip set for multimedia applications. *AT&T Technical Journal*, 72(1), 50–66.

ACM (2018). Fathers of the deep learning revolution receive acm turing award [Online]. Available: https://awards.acm.org/about/2018-turing.

Araki, T., Toyokura, M., Akiyama, T., Takeno, H., Wilson, B., and Aono, K. (1994). Video DSP architecture for MPEG2 codec. *Proceedings of ICASSP '94. IEEE International Conference on Acoustics, Speech and Signal Processing*, 2, II/417–II/420.

Blott, M., Preusser, T.B., Fraser, N.J., Gambardella, G., O'Brien, K., Umuroglu, Y., Leeser, M., and Vissers, K. (2018). Finn-R: An end-to-end deep-learning framework for fast exploration of quantized neural networks. *ACM Transactions on Reconfigurable Technology and Systems*, 11(3), 16:1–16:23 [Online]. Available: http://doi.acm.org/10.1145/3242897.

Chellappa, R., Bhattacharyya, S.S., Saha, S., Wolf, W., Aggarwal, G., Schlessman, J., and Kianzad, V. (2005). An architectural level design methodology for embedded face detection. *2005 Third IEEE/ACM/IFIP International Conference on Hardware/Software Codesign and System Synthesis (CODES+ISSS'05)*, pp. 136–141.

Chen, M.-J., Chen, L.-G., and Chiueh, T.-D. (1994). One-dimensional full search motion estimation algorithm for video coding. *IEEE Transactions on Circuits and Systems for Video Technology*, 4(5), 504–509.

Chen, Y., Chen, T., Xu, Z., Sun, N., and Temam, O. (2016). Diannao family: Energy-efficient hardware accelerators for machine learning. *Communications of the ACM*, 59(11), 105–112 [Online]. Available: http://doi.acm.org/10.1145/2996864.

Dutta, S., O'Connor, K.J., Wolf, W., and Wolfe, A. (1998). A design study of a 0.25-/spl mu/m video signal processor. *IEEE Transactions on Circuits and Systems for Video Technology*, 8(4), 501–519.

Dutta, S., Jensen, R., and Rieckmann, A. (2001). Viper: A multiprocessor soc for advanced set-top box and digital TV systems. *IEEE Design Test of Computers*, 18(5), 21–31.

Frumusanu, A. (2019a). The apple iPhone 11, 11 Pro and 11 Pro Max review: Performance, battery, and camera elevated [Online]. Available: https://www.anandtech.com/show/14892/the-apple-iphone-11-pro-and-max-review/2.

Frumusanu, A. (2019b). Samsung announces Exynos 990: 7nm EUV, M5, G77, LPDDR5 flagship SOC alongside Exynos 5123 5G modem [Online]. Available: https://www.anandtech.com/show/15021/samsung-announces-exynos-990-7nm-euv-m5-g77-lpddr5-and-5g-modem-flagship-soc.

Horprasert, T., Harwood, D., and Davis, L. (1999). A statistical approach for real-time robust background subtraction and shadow detection. *Proceedings of the IEEE ICCV'99 FRAME-RATE Workshop*, session 2.

Huang, F., Huang, S., Ker, J., and Chen, Y. (2012). High-performance sift hardware accelerator for real-time image feature extraction. *IEEE Transactions on Circuits and Systems for Video Technology*, 22(3), 340–351.

Inoue, T., Goto, J., Yamashina, M., Suzuki, K., Nomura, M., Koseki, Y., Kimura, T., Atsumo, T., Motomura, M., Shih, B.S., Horinchi, T., Hamatake, N., Kumagai, K., Enomoto, T., Yamada, H., and Takada, M. (1993). A 300-MHz 16-b BiCMOS video signal processor. *IEEE Journal of Solid-State Circuits*, 28(12), 1321–1330.

Instruments, T. (1995). TMS320C80 (MVP) parallel processor. Technical Report SPRU110A.

Instruments, T. (2010). TMS320DM6446 digital media system-on-chip. Technical Report SPRS283H.

Krizhevsky, A., Sutskever, I., and Hinton, G.E. (2012). Imagenet classification with deep convolutional neural networks. *Proceedings of the 25th International Conference on Neural Information Processing Systems – Volume 1, NIPS'12*, Curran Associates Inc., Red Hook, NY, USA, pp. 1097–1105.

Liu, B. and Zaccarin, A. (1993). New fast algorithms for the estimation of block motion vectors. *IEEE Transactions on Circuits and Systems for Video Technology*, 3(2), 148–157.

Lowe, D.G. (1999). Object recognition from local scale-invariant features. *Proceedings of the Seventh IEEE International Conference on Computer Vision*, vol. 2, pp. 1150–1157.

Lucas, B.D. and Kanade, T. (1981). An iterative image registration technique with an application to stereo vision. *International Joint Conference on Artificial Intelligence*, pp. 674–679.

Miyake, J., Urano, M., Inoue, G., Yano, J., Tsubata, S., Nishiyama, T., and Yamaguchi, S. (1998). Architecture of 23gops video signal processor with programmable systolic array. *IEEE Transactions on Circuits and Systems II: Analog and Digital Signal Processing*, 45(9), 1272–1278.

Netravali, A.N. and Robbins, J.D. (1979). Motion-compensated television coding: Part I. *The Bell System Technical Journal*, 58(3), 631–670.

Nomura, M., Suzuki, J., and Ono, S. (1989). Implementation of video codec with programmable parallel DSP. *1989 IEEE Global Telecommunications Conference and Exhibition "Communications Technology for the 1990s and Beyond"*, 2, 908–912.

Pirsch, P. and Gehrke, W. (1995). VLSI architectures for video signal processing. *Fifth International Conference on Image Processing and its Applications*, pp. 6–10.

Prost-Boucle, A., Bourge, A., and Pétrot, F. (2018). High-efficiency convolutional ternary neural networks with custom adder trees and weight compression. *ACM Transactions on Reconfigurable Technology and Systems*, 11(3), 15:1–15:24. [Online]. Available: http://doi.acm.org/10.1145/3270764.

Schlessman, J. and Wolf, M. (2015). Tailoring design for embedded computer vision applications. *Computer*, 48(5), 58–62.

Schlessman, J., Saha, S., Wolf, W., and Bhattacharya, S.S. (2005). An extended motion-estimation architecture applied to shape recognition. *2005 IEEE International Conference on Multimedia and Expo*, pp. 1504–1507.

Schlessman, J., Lodato, M., Ozer, B., and Wolf, W. (2007). Heterogeneous MPSOC architectures for embedded computer vision. *2007 IEEE International Conference on Multimedia and Expo*, pp. 1870–1873.

Sen, M., Corretjer, I., Haim, F., Saha, S., Bhattacharyya, S.S., Schlessman, J., and Wolf, W. (2005). Computer vision on fpgas: Design methodology and its application to gesture recognition. *2005 IEEE Computer Society Conference on Computer Vision and Pattern Recognition (CVPR'05) – Workshops*, pp. 133–133.

Shilov, A. (2019). Samsung kicks off mass production of AI chip for Baidu: 260 tops at 150 W [Online]. Available: https://www.anandtech.com/show/15247/samsung-kicks-off-mass-production-of-ai-chip-for-baidu-260-tops-at-150-w.

Slavenburg, G.A., Rathnam, S., and Dijkstra, H. (1996). The trimedia TM-1 PCI VLIW media processor. *Eighth Hot Chips Symposium*. Presentation.

Systems, C. (2019). Wafer-scale deep learning. *First Hot Chips Symposium*. Presentation.

Tesla (2017). NVIDIA Tesla V100 GPU architecture: The world's most advanced data center GPU. Technical Report WP-08608-001_v1.1.

Theocharides, T., Link, G., Vijaykrishnan, N., Irwin, M.J., and Wolf, W. (2004). Embedded hardware face detection. *17th International Conference on VLSI Design, Proceedings*, pp. 133–138.

Tomasi, C. and Kanade, T. (1991). Detection and tracking of point features. Technical Report CMU-CS-91-132, Carnegie Mellon University.

Wolf, W., Ozer, B., and Lv, T. (2002). Smart cameras as embedded systems. *Computer*, 35(9), 48–53.

Wolf, W., Jerraya, A.A., and Martin, G. (2008). Multiprocessor system-on-chip (MPSoC) technology. *IEEE Transactions on Computer-Aided Design of Integrated Circuits and Systems*, 27(10), 1701–1713.

Wren, C.R., Azarbayejani, A., Darrell, T., and Pentland, A. (1997). Pfinder: Real-time tracking of the human body. *IEEE Transactions on Pattern Analysis and Machine Intelligence*, 19(7), 780–785.

Xu, J., Wolf, W., Henkel, J., Chakradhar, S., and Lv, T. (2004). A case study in networks-on-chip design for embedded video. *Design, Automation and Test in Europe Conference and Exhibition, 2004, Proceedings*, 2, 770–775.

Yang, K.-M., Sun, M.-T., and Wu, L. (1989). A family of VLSI designs for the motion compensation block-matching algorithm. *IEEE Transactions on Circuits and Systems*, 36(10), 1317–1325.

Yang, S., Wolf, W., and Vijaykrishnan, N. (2005). Power and performance analysis of motion estimation based on hardware and software realizations. *IEEE Transactions on Computers*, 54(6), 714–726.

10

Software Compilation and Optimization Techniques for Heterogeneous Multi-core Platforms

Weihua SHENG[1], Jeronimo CASTRILLON[2] and Rainer LEUPERS[3]

[1] *Hong Kong Research Center, Huawei, Hong Kong, China*
[2] *Chair for Compiler Construction, TU Dresden, Germany*
[3] *Institute for Communication Technologies and Embedded Systems, RWTH Aachen University, Germany*

This chapter addresses the challenges associated with compilation and optimization techniques for heterogeneous multi-core computing systems in the embedded industry. Wireless terminals and modems are typical examples of such systems, which demand high performance and energy efficiency at the same time. To fully exploit the computing power of those systems, the existing compiler technology for single processor systems, does not suit the need and scale for multi-core architectures anymore. The authors have applied a systematic approach to tackle the problems of application modeling, source-to-source compilation, flexible compiler infrastructure construction and software distribution for multi-core architectures from a practical perspective. Several real-world multi-core platforms, as well as, system-level virtual platforms have been successfully used to demonstrate the achievable speed-ups and versatility of the compilation and optimization techniques developed in this work.

For a color version of all figures in this book, see www.iste.co.uk/andrade/multi2.zip.

Multi-Processor System-on-Chip 2 – Applications,
coordinated by Liliana ANDRADE and Frédéric ROUSSEAU.

10.1. Introduction

In the HPC (High-Performance Computing) industry, computer architecture evolution is predominantly driven by the increasing demand for superior performance. The trend of scaling up the frequency and number of transistors continued as long as possible, until the physical limits were reached and the power consumption became unmanageable, and CPU performance growth hit a wall around 2003. Between 2003 and 2005, multi-core architectures were first introduced in mainstream computing products, such as desktop PCs and high-end servers, by semiconductor vendors, for enhanced computing capabilities. The embedded industry followed suit, particularly adopting those multi-cores that consist of different types of processors (a.k.a. heterogeneous multi-cores).

Take the mobile phone as an example. Figure 10.1 shows the ongoing mobile traffic growth in western Europe, with projections until 2021. In 2016, western Europe's mobile data traffic amounted to 736,377 terabytes per month (Cisco Systems 2017). By 2021, the western European traffic from mobile devices (smartphones, laptops, tables, M2M connections, etc.) is projected to reach 4,189,615 terabytes per month. This projected compound annual growth rate (CAGR) amounts to 42%.

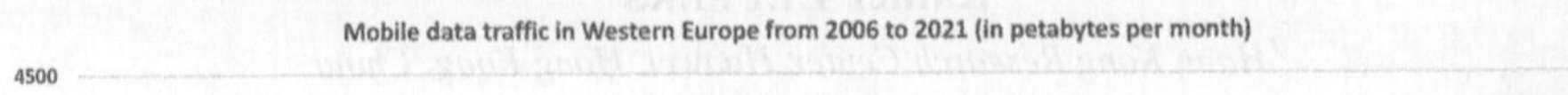

Figure 10.1. *Mobile data traffic in western Europe from 2006 to 2021 (2019-2021 are projections), adapted from Page* et al. *(2010) and Cisco Systems (2017)*

High-end smartphones, nowadays, feature high-definition video playback, fully fledged Internet capabilities and multi-standard radio protocols. The list of integrated functions demanded by consumers continues to grow. The workload of a typical 4G smartphone had increased to 100 giga operations per second (GOPS) per Watt in 2010, including 40 GOPS for radio, 20 GOPS for media processing and 6.5 GOPS for graphics. The workload continues to increase at a steady rate, roughly by an order of magnitude every five years (van Berkel 2009). Multi-processing was introduced to mobile phone technology to address this requirement shortly after the shift in the HPC industry. The International Technology Roadmap for Semiconductors (ITRS) has recently predicted an exponential growth in the number of processing elements that consumer portable products will contain, which is estimated to reach almost 6000 cores by 2026 (as illustrated in Figure 10.2).

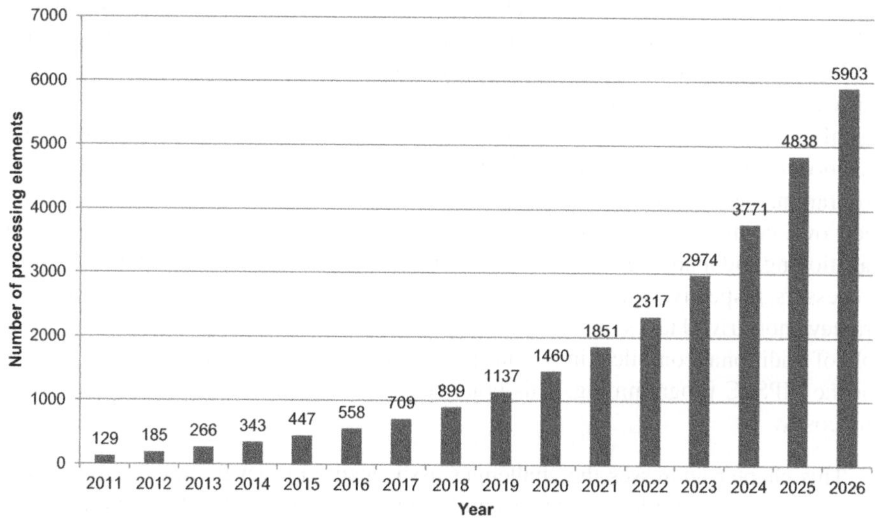

Figure 10.2. *SOC consumer portable design trends (adapted from International Technology Roadmap for Semiconductors (2011)). The processing elements are referred to cores or parallel execution units in CPUs, GPUs, accelerators, etc.*

Compared to other domains, embedded systems have much more stringent requirements in terms of power/energy efficiency. Today's mobile phones are expected to run on a single battery for hours, to days, without requiring recharging. As mentioned earlier, the power budget for the digital workload of a 4G smartphone (100 GOPS) is around 1 W. This is significantly more power efficient than most modern PCs. The breakdown analysis of the workload versus power consumption has shown that even homogeneous multi-processing, is realistically insufficient to meet this requirement (van Berkel 2009). The only feasible approach is to adopt heterogeneous multi-core architectures that consist of specialized programmable processors and/or hardware accelerators; these offer a superior solution for the trade-off between performance and power.

Unfortunately, the flourishing of multi-core architectures led by semiconductor companies has left several unresolved challenges from the perspective of software development.

To illustrate the current practice of MPSoC programming, Figure 10.3 shows a comparison of programming flows for uni-processor systems and for MPSoCs. In the uni-processor flow, software programmers follow the sequential programming model (C being the most popular language) and rely on the compilers to generate the target-specific code correctly and optimally, as shown in Figure 10.3(a). Software programmers focus on structuring algorithms correctly and are shielded from low-level architectural details, by using a vendor compiler. This has been a successful practice in recent decades prior to the introduction of MPSoCs, thanks to the pivotal role of compilers. However, the traditional compiler technology does not scale for MPSoCs. Figure 10.3(b) demonstrates the current problematic programming flow for MPSoCs, which results in low software productivity. Applications must first be partitioned into parallel tasks, followed by spatial and temporal mapping of those partitioned tasks, onto the MPSoC processing elements. As explained earlier, the programmable processors in heterogeneous MPSoCs, nowadays, often come with their own compilers and have their own software stacks (API, OS). Therefore, after partitioning and mapping, the correct code must be generated for each of the individual processors, respectively, to be further compiled. Compared to the uni-processor, these are new, non-trivial tasks that are the programmers' responsibility. Unlike the pivotal role of traditional compilers in the uni-processor flow, there is little compiler support for the MPSoC programming. This practice is currently labor-intensive, error-prone and costly.

This chapter addresses the challenges associated with developing compilers and their optimizations for heterogeneous multi-cores for software development, focusing on embedded systems and using dataflow models[1]. This work has been developed as part of the MPSoC Application Programming Studio (MAPS) research project, at the Institute for Communication Technologies and Embedded Systems (ICE) of RWTH Aachen University, which provided the technology foundation of the software tool provider Silexica Inc.

The remainder of the chapter is organized as follows. The dataflow modeling (especially process networks) is discussed in section 10.2. A clean, lightweight C language extension, CPN (i.e. C for process networks), is defined to capture streaming programming models. A retargetable source-to-source compiler has been developed to provide the key capabilities that facilitate the construction of compiler infrastructures for real-world, complex multi-core architectures. Their details are explained in

1 Sequential code partitioning is also an important and essential part of this work. Due to the chapter length constraint, we omit the details here. Refer to Leupers and Castrillon (2010), Castrillon *et al.* (2011) and Castrillon Mazo (2013).

section 10.3. Some optimization techniques for software distribution are explained in section 10.4. Our experimental results are presented in section 10.5. The chapter ends with a short summary in section 10.6.

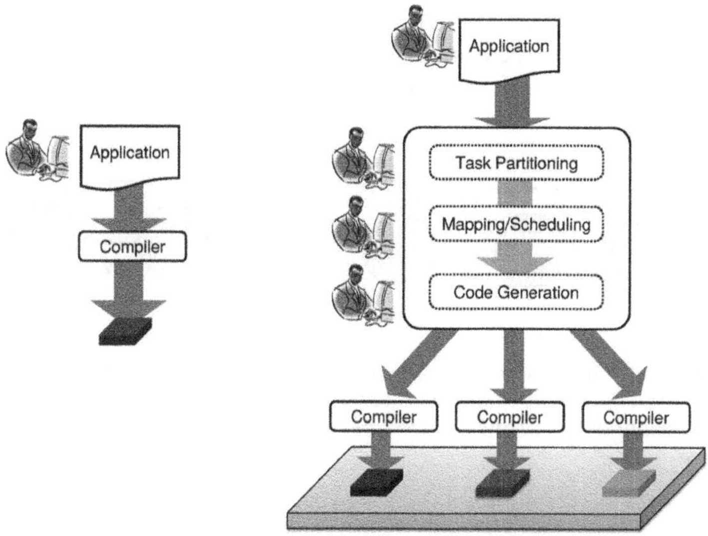

(a) (b)

Figure 10.3. *Programming flow. (a) Uni-processor: compilers perform an end-to-end translation from parsing source code, target independent optimization and target-dependent optimization to final binary generation. (b) MPSoC: no compilation framework for MPSoCs is in place. The process (task partitioning, mapping/scheduling and code generation) is manual*

10.2. Dataflow modeling

10.2.1. *General concepts*

Dataflow models subsume a family of models of computation, where parallel entities are only allowed to communicate over dedicated channels (Ptolemaeus 2014). Intuitively, in a dataflow model, an application is represented as a graph, where nodes represent computation and edges represent first-in first-out (FIFO) communication buffers. Dataflow models differ from each other in the way computation can be triggered upon conditions in incoming edges. For instance, in synchronous dataflow (SDF) (Lee and Messerschmitt 1987), a prominent dataflow model, nodes have a fixed condition upon which they can be executed (or *fired*). Additionally, once a node is executed, it always produces a fixed amount of data items called tokens. In the example in Figure 10.4(a), node *a2* requires four data items in its input channel and, when executed, it consumes the tokens and produces a single data item in its output. Such a node can, for example, model an averaging task that takes four integers and returns the average as the floating number. A model, as in the example, enables automatic reasoning about how such an application may make progress. For example, given that two tokens are available in *e3*, node *a1* can be executed twice, each time producing two tokens on *e1*. If extended with estimates of the duration of each node execution, a schedule can be constructed to understand the earliest time at which *a2* can be executed. The ability to construct efficient schedules at compile time made SDFs quite appealing for modeling real-time applications.

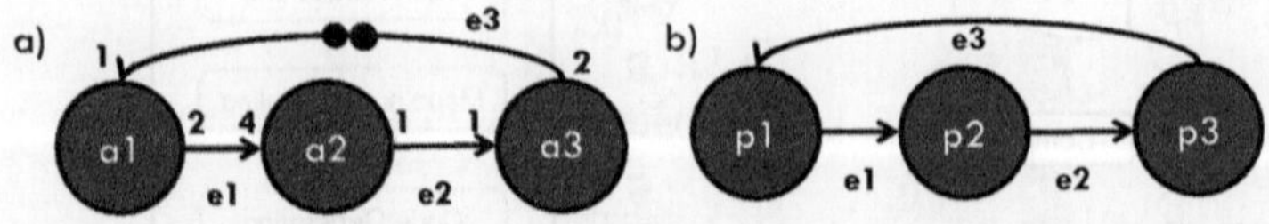

Figure 10.4. *a) Example of an SDF graph. Actors have fixed consumption and production rates. Initial tokens are represented as dots on the channels. b) Example of a process network. The abstract specification expresses that nodes can communicate if there is a channel that connects them, but does not specify how this communication takes place*

SDFs are simple in that they can only model static behavior. Over time, models have been extended to describe applications with more dynamic behavior, for example, cyclo-static SDF (CSDF) (Bilsen *et al.* 1996) and variable-rate phased dataflow (Wiggers 2009) (see also (Stuijk *et al.* 2011), for an attempt to compare different models). The popularity of these models for embedded software (Sriram and Bhattacharyya 2009; Bhattacharyya *et al.* 2010), motivated research on frameworks for the analysis and synthesis of dataflow models (DOL (Thiele *et al.* 2007; Giannopoulou *et al.* 2016), CAL (Eker and Janneck 2003; Brunet 2015), MAPS (Leupers and Castrillon 2010; Leupers *et al.* 2017), SystemCoDesigner (Haubelt *et al.* 2007; Keinert *et al.* 2009), PREESM (Pelcat *et al.* 2014) or Daedalus (Nikolov 2009; Bamakhrama *et al.* 2012) to name a few).

10.2.2. *Process networks*

Process networks are a model of computation in which applications are also represented as a graph with FIFO communication channels, similar to dataflow models. In contrast to dataflow models, there are no triggering conditions specified in the input channels of a node in a process network, i.e. no *firing semantics*. Instead, the graph representation only specifies that nodes (aka processes) can communicate if a channel connects them. In Figure 10.4(b), for instance, processes *p1* and *p2* could execute in parallel for a long time before any actual data exchange takes place. A process network can model not only SDF, CSDF, and other simple dataflow graphs, but also more complex communication behavior. In particular, in a process network, it is possible to model channel accesses that depend on concrete values of incoming data.

Process networks are more expressive than many dataflow models, in the sense that they can model a larger set of applications. Since every process can be thought of as an independent thread, questions about termination or scheduling are undecidable. An interesting property of Kahn process networks (KPNs) is that their execution is *determinate* (Kahn 1974), i.e. provided with the same input, the history of tokens in all the channels is independent of the schedule. This is enabled by the fact that processes are not allowed to peek into channels before reading, but would immediately block when attempting to read from an empty channel. This is typically called *blocking-read semantics*. Intuitively, whether peeking into a channel is successful, depends on the time of arrival of tokens, which does depend on the schedule. Determinism is an important property that enables automatic optimization of process networks, without modifying the application behavior.

10.2.3. *C for process networks*

MAPS defines a clean, lightweight C language extension called CPN (C for Process Networks) to capture streaming models. A minimum set of new keywords is added to the C language to describe processes and channels. A language extension approach allows the user to specify the semantics of process networks at a high level, for example, containing enough structural information about the channel accesses. This first enables retargetability towards typical embedded MPSoCs, where processing elements have different APIs and specific low-level primitives (that often cannot be abstracted by a common API). Second, programs are portable and the semantic analysis offers abundant opportunities for code transformations and optimizations (e.g. process fusion and fission). Although C is not an ideal vehicle in which to carry the concurrency specification, this design compromise is made considering the large C legacy code base and popularity of C in the embedded industry.

10.2.3.1. *Channels*

Channel declaration in CPN is similar to declaring a global variable in C, with an additional keyword `__PNchannel`. Examples can be found in Figure 10.5. Elementary C types, such as `int`, `char` and `float`, and enumerations are valid channel types. Structures, unions and arrays of valid channel types are valid, too.

A channel is, by default, empty at the beginning of the execution. If initial tokens are needed on the channels, for example, to avoid deadlocks, CPN also supports having initial tokens in channels by specifying initializers in the channel declaration (channel `A` at line 7 in Figure 10.5).

```
typedef struct { int i; double d; } my_struct_t;
typedef union { float f; short s[4]; } my_union_t;

__PNchannel char B[3][3];
__PNchannel my_struct_t C;
__PNchannel my_union_t D;
__PNchannel int A = {1, 2, 3}; /* Initial channel tokens */
```

Figure 10.5. *CPN example code: channel declaration*

10.2.3.2. *Processes*

Similar to C++ templates, the concept of process templates is used in CPN for code reuse. A process template describes the functionality of a process and the channels this process needs to access (either read or write). Processes are always created as instances of process templates. The code's readability and conciseness is improved, for example, when multiple processes in a network share the same functionality.

An example of a KPN process template is shown in Figure 10.6. It describes the functionality of run-length decoding (RLD). Run-length encoding (RLE) is a simple data compression technique used in fax machines, for example. Data are encoded into a stream of duos, i.e. the number of appearances and the data element itself. For example, the original data `AAAABBCCCCCDDD` is compressed into `4A2B5C3D`. RLD is the inverse of RLE. A KPN process template (`__PNkpn`) with the identifier `RLD`, is shown in Figure 10.6. It reads integers from its input channel `EncIn` and outputs integers to the channel `DecOut` indicated by keywords `__PNin` and `__PNout`, respectively. The body of a KPN process template can contain arbitrary code, which is allowed to access input and output channels at any point of its control flow.

The access to a channel is always explicitly made via `__PNin` or `__PNout` in the body for KPN-like processes, as the code generator cannot know if two consecutive accesses to the same channel intend to access the same channel item (token), or the next channel item in the general case. Those statements enable access to the next data tokens (read) or free entries (write) in the channel. The code inside the body of the `__PNin` or `__PNout` statement can access those items such as a local variable. At lines 6 and 7 of Figure 10.6, first, a token from the input channel `EncIn` is read and assigned to a local variable `count`. Next, the data are decoded by writing the encoded data into the output channel `DecOut` using in the loop of lines 9-11. Both the `__PNin` and the `__PNout` statements have blocking semantics, i.e. they will suspend process execution until the channel contains enough data tokens or free entries.

The explicit channel access annotation looks somewhat unnecessary in the first place. A brief explanation is provided here to illustrate why such an explicit annotation is required, using Figure 10.6 as an example. If the example were without `__PNin` and `__PNout` statements, the accesses to channel `EncIn` at lines 7 and 11 would become ambiguous. It is not possible for the code generator to reason that, in the case of line 7, a new token from the channel `EncIn` is read, while in the case of line 11, a new token is read from the channel `EncIn`, only at the first loop iteration and will be reused during the rest of the loop (lines 9-11). Any form of dependence analysis would not be sufficient to solve the problem because it requires the understanding of the application behavior. Therefore, the functionality of RLD could not be correctly specified. The code generator requires a clear, unambiguous indication as to whether or not to fetch tokens from a channel (or write tokens to a channel).

```
1  /* Run Length Decoding, e.g. 4A2B5C3D -> AAAABBCCCCCDDD */
2  __PNkpn RLD __PNin(int EncIn) __PNout(int DecOut)
3  {
4      int count, i;
5      while (1) {
6        __PNin(EncIn) /* read a token (# of appearances) from EncIn,
                         e.g. 4 */
7          count = EncIn;
8        __PNin(EncIn) /* read a token (data itself) from EncIn,
                         e.g. A */
9          for (i = 0; i < count; ++i) /* write data to DecOut,
                                         e.g 4 times of A */
10             __PNout(DecOut)
11                 DecOut = EncIn;
12     }
13 }
```

Figure 10.6. *CPN example code: KPN process template (run-length decoding)*

As the SDF frequently appears in streaming applications, a shortcut is provided in CPN, to simplify the program representation. The example in Figure 10.7 defines an SDF process template (`__PNsdf`) for the functionality of an adder, reading integers from input channels `a` and `b` and writing integers to channel `sum`. The `__PNloop` statement resembles the infinite loop of an SDF process. Its body contains all the code to be executed between reading from all input channels and writing to all output channels. SDF-like processes implicitly access all their input and output channels in the `__PNloop` statement (line 3 in Figure 10.7). The number of tokens to access in every iteration is given in the `__PNin` and `__PNout` clauses of the SDF template header. If this number is not specified, the default value 1 is used.

Processes can be instantiated from previously defined process templates using `__PNprocess`. Figure 10.8 creates processes from the process templates defined above and connects channels to them.

10.2.3.3. *Parallelism types*

CPN is designed to describe streaming applications and, more importantly, specify the parallelism in those programs explicitly. In this section, the parallelism types that are most commonly seen in embedded streaming applications are introduced first, and we demonstrate that CPN is able to express all of those parallelism types succinctly.

```
__PNsdf Add __PNin(int a, int b) __PNout(int sum) {
  /* initialization code could be placed here */
  __PNloop { /* infinite loop */
    /* a and b are read from the channel implicitly */
    sum = a + b; /* channel variables are accessible in C code */
    /* sum is written to the channel implicitly */
  }
}
```

Figure 10.7. *CPN example code: SDF process template (adder)*

```
__PNchannel int decoder1_input = {4,'A',2,'B',5,'C',3,'D'};
__PNchannel int decoder2_input = {3,'E',5,'F',4,'G',2,'H'};

__PNchannel int decoder1_output, decoder2_output;
__PNchannel int add_output;

__PNprocess decoder1 = RLD __PNin(decoder1_input)
    __PNout(decoder1_output);
__PNprocess decoder2 = RLD __PNin(decoder2_input)
    __PNout(decoder2_output);
__PNprocess add = Add __PNin(decoder1_output, decoder2_output)
    __PNout(add_output);
```

Figure 10.8. *CPN example code: process instantiation*

Many different types of parallelism exist in real-world applications. In the embedded domain that is the focus of this chapter, the most prominent types of coarse or macroscopic parallelism are data-level, task-level and pipeline-level. These are illustrated in Figure 10.9 with descriptions below.

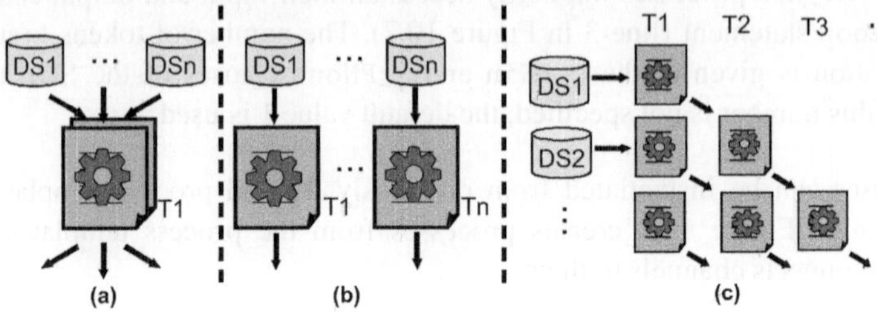

Figure 10.9. *Types of parallelism: (a) DLP, (b) TLP and (c) PLP (DS stands for data set and T stands for task)*

– Data-level parallelism (DLP): DLP represents a typical scenario where the same computational task is carried out on several disjoint datasets, as illustrated in Figure 10.9(a). DLP can be considered as a generalization of single instruction, multiple data (SIMD). DLP exists abundantly, for example, in multimedia applications, where a decoding task performs the same operations on different portions of an image or video (e.g. Macroblocks in H.264). An example code to describe the DLP is shown in Figure 10.10. As with other programming models such as OpenMP, where data parallelism is defined implicitly, we have worked on extensions that allow us to define data-parallel processes and methodologies, to change the amount of parallelism dynamically (Khasanov *et al.* 2018).

```
__PNchannel int DS1, DS2, DSN;
__PNchannel int Result1, Result2, ResultN;
__PNprocess T_1 = SameTaskT1 __PNin(DS1) __PNout(Result1);
__PNprocess T_2 = SameTaskT1 __PNin(DS2) __PNout(Result2);
__PNprocess T_N = SameTaskT1 __PNin(DSN) __PNout(ResultN);
```

Figure 10.10. *CPN example code: DLP*

– Task (or functional)-level parallelism (TLP): It is common that in a computer program, different tasks can compute in parallel on different datasets. TLP is used to specify this kind of behavior, as shown in Figure 10.9(b). Tasks may have dependencies on one another, but once a task has its data ready, it can execute in parallel with the already-running tasks in the system. An example code to describe the TLP is shown in Figure 10.11.

```
__PNchannel int DS1, DS2, DSN;
__PNchannel int Result1, Result2, ResultN;
__PNprocess T_1 = Task1 __PNin(DS1) __PNout(Result1);
__PNprocess T_2 = Task2 __PNin(DS2) __PNout(Result2);
__PNprocess T_N = TaskN __PNin(DSN) __PNout(ResultN);
```

Figure 10.11. *CPN example code: TLP*

– Pipeline-level parallelism (PLP): In PLP, computation is broken into a sequence of tasks that are executed repetitively for different datasets, as shown in Figure 10.9(c). These tasks are also called pipeline stages. It follows the principle of pipelining to achieve a higher throughput. At a given time, different tasks of the original functionality are executed concurrently on different datasets. This type of parallelism is most common in the signal processing domain, where the throughput is a critical system performance indicator. An example code to describe the PLP is shown in Figure 10.12.

```
__PNchannel pixels DS, intermediate1, intermediate2,
  intermediate3, output;
__PNprocess source = ReadImage __PNout(DS);
__PNprocess T_1 = PipelineTask1 __PNin(DS)
   __PNout(intermediate1);
__PNprocess T_2 = PipelineTask2 __PNin(intermediate1)
   __PNout(intermediate2);
__PNprocess T_3 = PipelineTask3 __PNin(intermediate2)
   __PNout(intermediate3);
__PNprocess T_N = PipelineTaskN __PNin(intermediate3)
   __PNout(output);
__PNprocess sink = WriteImage __PNin(output);
```

Figure 10.12. *CPN example code: PLP*

10.3. Source-to-source-based compiler infrastructure

Once applications are written in CPN, a compiler framework is needed to take them as input, compile and generate the correct and optimized binary code for target MPSoC platforms. The complexity of MPSoCs nowadays requires a more flexible design in reality. In this section, the compiler design rationale is discussed, followed by the details of our implementation.

10.3.1. *Design rationale*

Many previous approaches for multi-processor systems have followed the principle of compiler design for uni-processors: the compiler works on input programs, builds an internal intermediate representation, performs optimizations and generates the target binary code. This black box monolithic design philosophy is:

1) the compiler generates the correct and optimized target code without (much) help from programmers;

2) the compiler has as much information about architectural details as possible, in order to perform target-specific code generation and optimization.

The essence of this design philosophy is to incorporate the complexity into a single tool (compiler) that has been developed by a small group of highly skilled experts. Therefore, programmers exhibit enhanced productivity in designing the software that assumes a straightforward programming model (sequential) and a simple, common memory model.

We studied and revisited this monolithic design approach used in the past few decades (shown in Figure 10.13) and our observations are as follows:

1) The complexity of hardware architectures grows so rapidly that a monolithic compiler is unable to keep up with it. For most uni-processors, such as RISCs,

compilers still manage to include architectural details. However, the gap between the architecture complexity and the amount of information that a single tool can incorporate, has grown wider: more complex architecture templates appear, such as (clustered-) VLIW DSPs, ASIPs (Application-Specific Instruction set Processors) and eventually multi-cores, which have multiple scalar processors on a single chip. The level of user intervention required in developing and using compilers, increases rapidly (shown in Figure 10.13), which offers further proof that the monolithic approach is reaching its feasibility limits.

2) Compilers usually take a significant amount of time to mature along with extensive investment. As already explained, heterogeneous MPSoC platforms will increasingly use individual IPs from different vendors. Therefore, a monolithic approach to multi-cores that generates the target-specific binary code (for different processors) directly, is not economically sound, since it does not leverage the existing C compilers for uni-processors.

3) Compiler usage scenarios are now more diverse, particularly in the embedded domain. Multiple optimization goals are commonly seen, such as the performance, the energy efficiency and the code size.

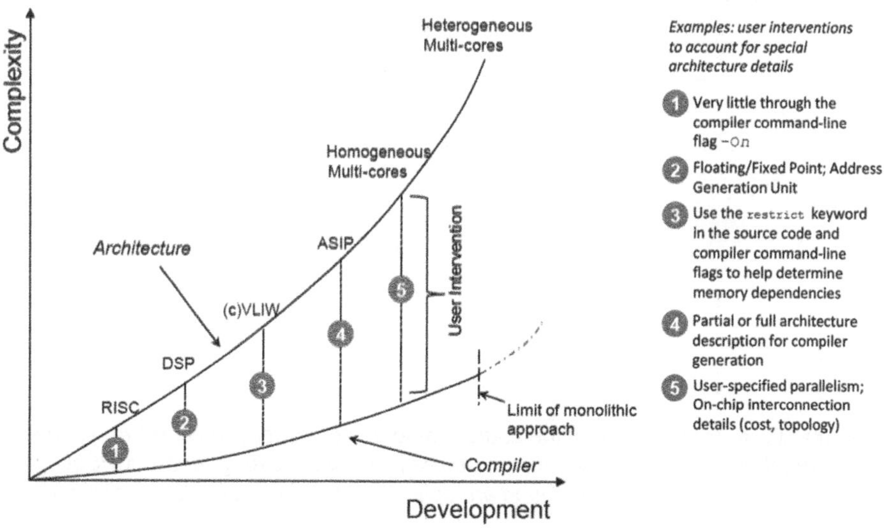

Figure 10.13. *Evolution of compiler development versus architecture development*

Moving along the trajectory shown in Figure 10.13, we argue that for heterogeneous multi-core systems, a new flexible, extensible compiler design is both desirable and required, as opposed to persisting with the monolithic approach. The user interventions to account for the new architecture trends will increase, including a user-defined mapping specification and objective functions for code generation and optimization.

10.3.2. *Implementation strategy*

The software architecture of the compiler framework developed to compile CPN programs to heterogeneous MPSoCs is introduced below. The complete compilation flow for a specific MPSoC is a tool framework consisting of many components. The core component cpn-cc, a source-to-source (CPN-to-C) compiler, is first elaborated, followed by a description of how the complete compilation framework can be constructed in a flexible, extensible manner guided by the user.

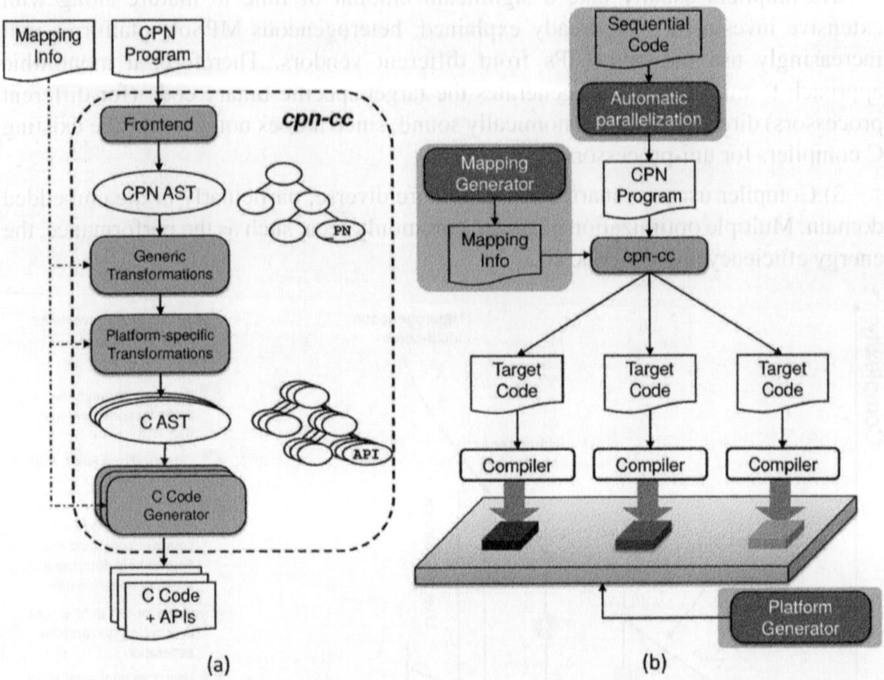

Figure 10.14. *(a) Structure of the source-to-source compiler cpn-cc. (b) An example of a complete compiler framework for a three-processor heterogeneous MPSoC: the parts with a shaded background are optional to the core compiler framework*

Instead of a monolithic approach, a source-to-source (CPN-to-C) compiler, cpn-cc, was developed as the core component in the framework. The cpn-cc is implemented based on Clang (Lattner 2008), the C frontend of the LLVM compiler infrastructure. Figure 10.14(a) shows a high-level, internal structure of cpn-cc which consists of the frontend, generic and platform-specific transformations, and a C code generator. Those components and some special considerations during the implementation are described below:

– Frontend: the frontend starts with C preprocessors to provide the functionalities, such as the inclusion of header files. After preprocessing, the source code is processed by the tokenizer, to which new CPN keywords were added, and the parser, which was extended with new grammar rules for CPN syntax elements, such as process templates and channel accesses. Then, the CPN-aware semantic analysis builds the abstract syntax tree (AST), as an intermediate representation in the memory for further processing. It contains all CPN language elements and constructs occurring in the source code.

– AST transformations: after preparation by the frontend, the AST transformations perform source-to-source translation. They are categorized into generic and platform-specific transformations, similar to the sequence of code optimizations in a classic uni-processor compiler (Aho *et al.* 2006). Generic transformations are platform-independent. One example is the transformation from SDF templates to KPNs, whereby SDF process templates are rewritten into KPNs with an endless loop, with explicit channel accesses in the behavior code. Platform-specific transformations replace the AST nodes of CPN constructs by C nodes, representing platform-specific API calls, for example FIFO primitives. Therefore, the result of the AST transformations will be a plain C code AST without any of the extensions introduced by CPN (but with target-specific API calls). Several ASTs are built in this phase, one for each processor on the target MPSoC platform.

– C Code Generator: as the last step of source-to-source translation, the AST Clang printer generates the C source code from the ASTs after transformations. In some cases, other additional auxiliary files, such as configurations and makefiles, also need to be generated, to be further compiled with the individual processor native C compilers.

Compared to other source-to-source translation tools that are not based on a formally built AST (e.g. textual replacements), our approach provides a cleaner and more powerful infrastructure for source code transformation based on the AST. The full semantic information allows code optimizations in a larger program context. For example, data type checking and variable manipulation can be carried out in the compilation, while other textual replacement approaches cannot do this reliably. The readability of the generated source code is also retained and the structure is close to the original code. However, compared to other approaches not based on an AST, more effort is needed to build a fully working compiler. The effort is well justified, in that cpn-cc is designed to be retargetable, and most cpn-cc components can be re-used for different target MPSoC platforms.

A complete compilation framework for heterogeneous MPSoCs can be built around the cpn-cc in a flexible and extensible manner, to suit different MPSoC target platforms in different scenarios. An example of such a framework for a specific MPSoC is shown in Figure 10.14(b):

– In addition to CPN programs that are considered as the functional specification, an important input provided by the user is the mapping info, which specifies the spatial mapping of processes onto processing elements and temporal order for execution. The mapping info, together with the CPN program, determines the performance of the application on the target hardware platform. Unlike the monolithic approach, which tries to automatically compute the mapping and scheduling, we have designed the mapping info as an input to the core compiler in order to keep a clean and lean software architecture. The mapping info can be provided by the user manually or can be generated by an external tool.

– Transformations are implemented in a modular way and can thus be customized for different target platforms in a plug-and-play fashion. A large portion of common transformations can be reused among many MPSoC targets, which significantly eases the retargeting process of the framework. Target-specific transformations can be developed to better exploit the specific hardware details.

– The compilation framework can be used in collaboration with other state-of-the-art tools, thanks to the extensible design. For example, an intelligent mapping info generator, referred to as mapping generator in Figure 10.14(b), can be easily integrated with the core compiler framework. In some embedded system design scenarios, the software development needs to proceed simultaneously with the architecture development, for example, using a platform generator (see Figure 10.14). Our work also facilitates the retargeting of the compiler framework to work with the electronic system level (ESL) design tools for early system-level design.

The main concept of this flexible and extensible design is to enable the re-use of components in constructing compiler frameworks for different MPSoC platforms. Therefore, retargetability can be achieved with minimal effort. Furthermore, the retargeting only needs to be done once for every new platform. As designing embedded systems today inevitably involves many other third-party and vendor tools, this design also makes it easy to collaborate with those tools, thus preserving existing software investments.

10.4. Software distribution

The MAPS framework was built as an interactive tool for the programmer to modify and optimize the application within the IDE. To further aid the programmer, MAPS also includes a set of configurable algorithms to automate different steps of the optimization process. The most notable process is the software distribution, which accounts for *mapping* computation (processes) onto the heterogeneous cores of the platform and communication (channels) to communication resources (e.g. communication APIs, memories and interconnect). Several works have addressed automatic software distribution in a similar way to MAPS, for performance, energy, thermal distribution among other optimization goals, for example Thiele *et al.* (2007), Marwedel *et al.* (2011), Quan and Pimentel (2014), Hascoët *et al.* (2017) and Das *et al.* (2015). Given the requirements imposed by the process network programming model, MAPS relies on the analysis of traces to determine static and hybrid mappings, as discussed in the following.

10.4.1. *KPN analysis*

As discussed in section 10.2, static dataflow programming models such as SDF expose the communication patterns in the access rates to channels. For this reason, most frameworks for the analysis of such graphs, consider actors as black boxes. KPN applications do not do so, so that communication patterns need to be analyzed by looking into the actual implementation of the behavior of processes. Consider the example in Figure 10.15. Given the hypothetical control flow of process $P1$, a tool can infer how many times the process writes to channel $C2$, before reading from channel $C1$ (determined by the variable `x` and the initial value of `i`). In MAPS, a *gray* model of processes is obtained by analyzing the processes in isolation and collecting traces of channel access events, as exemplified on the right side of Figure 10.15.

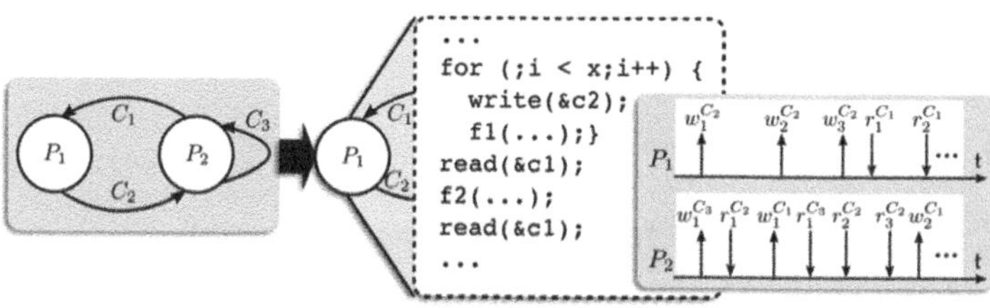

Figure 10.15. *Illustration of process traces. Example process network with a sample code of the logic of process $P1$, alongside a hypothetical trace of read and write events that may result from the process inner control flow*

Trace-based analysis is enabled by the deterministic nature of KPNs. That is, provided with the same input stimuli, every process will always produce the same history of tokens in the channels, independent of the schedule. MAPS leveraged this property to abstract processes, as trains of events with annotated execution costs, corresponding to the different core types of the platform (Castrillon *et al.* 2010). Technically, this was achieved via source code instrumentation and by using source code cost estimation tools (Gao *et al.* 2009; Eusse *et al.* 2014) whenever the target hardware was unavailable. Based on traces, algorithms could optimize the placement of events in the virtual execution time line, on the target heterogeneous systems, without changing the application semantics (see the next section). Naturally, different input stimuli would create different traces, each of which would have to be analyzed independently. Some recent efforts have tried to jointly optimize, for multiple execution traces of the same application, by clustering process behaviors (Goens and Castrillon 2015). Results are so far inconclusive.

10.4.2. *Static KPN mapping*

A static mapping is one that is computed at compile time and, typically, one in which compile-time decisions are not allowed to be overwritten at runtime. For the problem setup considered in MAPS, this boils down to (i) mapping computation, i.e. deciding which particular core executes which process, and (ii) mapping communication, i.e. deciding with which API over which hardware resources, a channel is to be implemented. The latter one allocates memory for communication channels and thus includes the process of buffer sizing, i.e. deciding what bounds are to be set on the logical FIFO buffers to reduce the chances of encountering deadlocks and improving throughput.

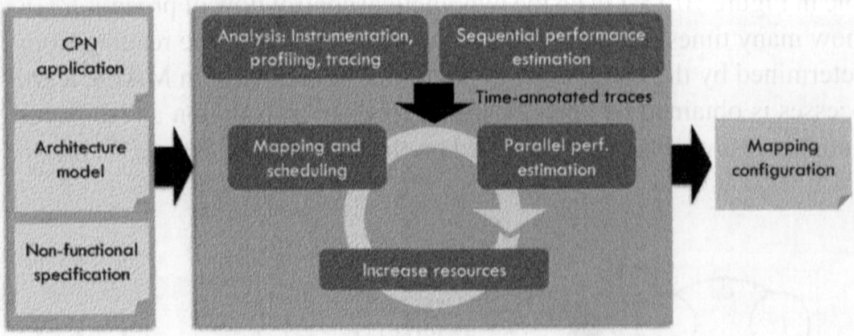

Figure 10.16. *Overview of the mapping flow*

Given the lack of information in KPN specification, most mapping approaches resort to metaheuristics, most prominently with genetic algorithms (Pimentel *et al.* 2006; Thiele *et al.* 2007; Nikolov 2009). MAPS aimed at leveraging the information in the traces to devise faster heuristics, to compute the mapping (Castrillon *et al.* 2013). The generic mapping flow within MAPS is shown in Figure 10.16. After analysis and sequential performance estimation, an iterative process is started that computes static mappings on an increasing set of allowed resources. By adding resources (cores and memory space), the tool can expose different trade-offs to the programmer and also attempt to find the minimal resource configuration that meets application constraints. Internally, each mapping configuration is simulated in a parallel performance estimation, that uses the traces collected during analysis, to emulate how the target system would execute the application.

Different heuristics are included in MAPS to compute the mapping of the application, given a set of resources. Among them, group-based mapping (GBM) is worth highlighting (Castrillon *et al.* 2012). This heuristic works on a graph representation of the traces, where dependencies between events are explicitly modeled with edges. Dependencies include traditional read-after-write dependencies which model the blocking read semantics of KPN applications. In the example trace in

Figure 10.15, the event $r_1^{C_2}$ of $P2$ can only succeed after process $P1$ has executed the corresponding write event ($w_1^{C_2}$). Blocking write semantics due to finite buffer sizes are also modeled via similar dependencies. In the example, assuming a buffer size of 2 for channel $C2$, would mean that the third write to the channel ($w_3^{C_2}$) can only execute after the consumer process $P2$ has read twice. More complex dependencies, for example, supporting events that write multiple tokens at once and more complex communication protocols are also supported (Odendahl *et al.* 2013). GBM uses this graph representation to compute mappings of traces onto cores in the platform, in a similar way to other heuristics for directed acyclic graphs. The key idea of GBM is to iteratively compute mappings of groups of application elements (processes or channels) to groups of platform resources (cores or memories). The priority in which groups are created and mapped is determined by the equivalent of the critical path, considering all possible mappings. By mapping groups, the heterogeneity of the platform is directly addressed. Concrete resources are decided in a final step. For further details, the reader is referred to Castrillon *et al.* (2012). For an analysis and comparison of heuristics and genetic algorithms, the reader is referred to Goens *et al.* (2016).

10.4.3. *Hybrid KPN mapping*

With embedded systems becoming more adaptive and running multiple workloads at once, fixed mapping approaches became outdated. Hybrid approaches refer to attempts to make mappings more flexible but still retain properties computed at compile time (predictable performance and energy consumption). Typically, this consists of producing multiple mapping configurations with different resource utilization at compile time and letting the runtime decide which variant to deploy (Schor *et al.* 2012; Quan and Pimentel 2013). In Quan and Pimentel (2014), for example, the authors propose a method in which pareto-optimal points are computed at compile time for different configurations, with an adaptable runtime presented in Quan and Pimentel (2016). These configurations are switched at runtime according to the given set of active applications. A similar strategy is used in Weichslgartner *et al.* (2016) and Schwarzer *et al.* (2018).

We have worked on identifying application and architecture symmetries, that allow us to define mapping equivalent classes (Goens *et al.* 2017b). Symmetries enable transformations by the compiler or the runtime that preserve design trade-offs, so that properties predicted at compile time are preserved at runtime. By making the runtime system aware of such transformations, we demonstrated better time predictability on off-the-shelf multi-core systems (Goens *et al.* 2017a). Our runtime featured up to 510× less performance variability and 83× less energy variability when compared to the default Linux scheduler on an Odroid XU4. We are currently working on joint application optimization, proactively looking into future workload change (Khasanov and Castrillon 2020).

10.5. Results

A tooling infrastructure must be validated in real use cases, which are presented below as experimental results. We have evaluated this design in:

– illustrative uses of MAPS for real-word MPSoC platforms and other scenarios; and

– how retargetability of MAPS for different MPSoC platforms is achieved.

10.5.1. *Applications and experiences*

MAPS, as an MPSoC compiler infrastructure, has been applied and retargeted to a number of MPSoC platforms, to demonstrate its feasibility and value in automating the compilation process. Those platforms include commercial platforms that are available in silicon, such as, TI OMAP3530 and TI C6678. Virtual prototypes of MPSoC platforms (e.g. Synopsys MCO (Schirrmeister and Sheridan 2011)) are also supported, and other multi-core platforms (mostly x86 or ARM-based) where a Pthreads environment is available. First, we briefly describe these platforms below:

– TI OMAP3530 was announced as a multimedia processor by TI in 2009 and is shown in Figure 10.17(a). It consists of two processors, an ARM Cortex A8 CPU running at 550 MHz and a TI C64x DSP clocked at 400 MHz. The hardware evaluation board used in this work possesses 128 Mbyte DDR SDRAM at 166 MHz, shared between the ARM and the DSP. Software-wise, Linux is used as OS on the ARM side. The Pthread library and the POSIX IPC (Inter-Process Communication) are available for realizing the thread management and communications. On the DSP side, a proprietary lightweight multitasking operating system, called DSP/BIOS (Dart 2001), is provided by TI. The communication between ARM and DSP is handled by TI's DSP/Link layer. Figure 10.17(b) gives an overview of the software stack for the OMAP platform. General TI OMAP programming strategies are discussed in Kloss (2003). The rule of thumb is to map signal processing tasks to DSP and control flow processing to ARM. However, no systematic methodology or tooling support is available.

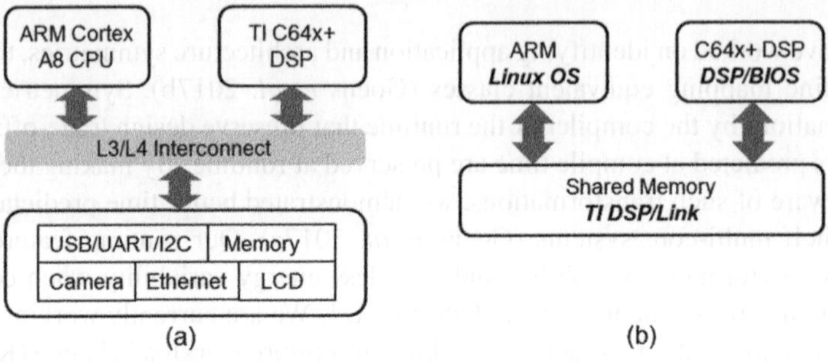

Figure 10.17. *(a) OMAP3530 block diagram. (b) Overview of TI OMAP software*

– TI C6678 (Texas Instruments 2011) is based on TI's KeyStone I multi-core architecture. It integrates eight C66x DSPs shown in Figure 10.18. Besides the local L1/L2 memory for each DSP and global shared memory, the C6678 platform is featured with the multi-core navigator that consists of more than 8,000 queues for direct communications between processor cores. This feature is particularly desirable for streaming programs. A software development toolchain (compiler, linker, etc.) is available for the DSP from TI. However, no platform-wide compilation framework is available.

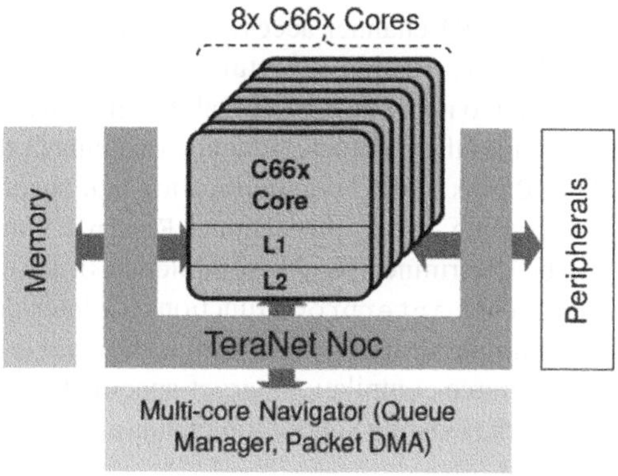

Figure 10.18. *TI C6678 block diagram*

– Synopsys MCO (Multi-Core Optimization) technology is a SystemC-based virtual platform solution for hardware/software co-development in the early system design phase. Multi-core platforms can be modeled using a high-level abstraction of a processing element, named virtual processing unit (VPU) (Kempf *et al.* 2005). The VPUs on the platform are able to execute SystemC modules that model application tasks. Those tasks contain C code that can be extended with explicit timing annotations and with communication directives, to access the communication infrastructure of the modeled hardware platform. An automatic compilation framework is desired for multi-core platforms in MCO, to enable seamless hardware/software co-development. MAPS has been retargeted to support MCO so that system architects can quickly evaluate software design choices, on a high-level virtual platform.

– Pthreads is a widely known parallel programming API and is widely available on multi-core platforms (e.g. x86 or ARM-based). MAPS has been retargeted to generate streaming programs using the Pthreads API to realize concurrent processes with FIFO communications. This expands the platforms that MAPS supports to virtually all platforms that are capable of running Pthreads. Alternatively, this option is also valuable when programmers perform functional verification on the host.

We have successfully applied the MAPS compiler infrastructure for these MPSoC platforms. Although there are some differences in different MAPS instances for these platforms, the basic principle and flow apply to all. As an example, the case of using MAPS for the TI OMAP3530 is introduced below, as shown in Figure 10.19. The example from Figure 10.8 is used as an input program for the cpn-cc compiler. A mapping info file is required for cpn-cc to perform source-to-source compilation, for example, the spatial mapping of processes to processors of OMAP3530. The cpn-cc compiler builds and transforms the AST to replace `__PN` nodes by OMAP3530-specific APIs. On the right side of Figure 10.19, a simplified code excerpt of the transformation results for FIFO channel accesses is shown. The generated code is both editable and readable, thus creating opportunities for performance fine-tuning by programmers. It is important to note that it is critical that all transformations take place at the AST level, that provides the complete semantic contents of the source code. For instance, the channel `DecOut` is a FIFO communication that occurs between different processors under this mapping, while the channel `EncIn` communication is local. This information must be determined by the compiler in order to select the correct target-specific API: in this case, `interproc_` functions for inter-processor FIFOs and `local_` for local FIFOs. Simple textual replacement techniques are insufficient in this case. After the source-to-source compilation, target-specific C files are generated for the ARM and the DSP and the existing vendor toolchains are re-used. The compilation flow is thus fully automated.

Two case studies of applications are presented here by using MAPS on the TI multi-DSP platform C6678 for completeness. Two signal processing benchmarks were selected: a digital audio filter and an airborne radar application.

The digital audio application implements fast convolution filtering using the fast Fourier Transform (FFT) and inverse FFT and it performs low-pass filtering on a stereo audio signal. Two parallelized versions of the application are written in CPN programming model to explore parallelism: (a) exploiting parallel processing on the left and right channels of the input signal simultaneously and (b) further parallelization by splitting 1024-point Fourier transforms into two 512-point blocks. Figure 10.20 shows graphical representations of both versions of the digital filter. The benchmark existed initially as a sequential C code. The conversion from sequential C code to parallel CPN versions took around half a day to complete, thanks to the CPN being close to C. As the CPN language separates the functional specification of processes and overall topology specification, the two versions differ only by a few lines, in that the further parallelization mainly modifies the topology.

To evaluate benchmarks on the C6678 platform, we used a mapping heuristic, group-based mapping (GBM), explained in section 10.4, to generate spatial mapping for benchmarks. The GBM takes constraints, for example the available number of processors, as inputs to compute mapping. Figure 10.21 shows the performance results when the constraint on the number of available processors increases from 1 to 8 for the audio filter. Each data point corresponds to the runtime result of the generated mapping, which is measured on the hardware board. It is evident that the speed-ups achieved increase, when the number of available processors rises. The higher number of parallel processes in Audio-2 gives better results than that in Audio-1, in that eight cores on the C6678 platform can be better used by Audio-2.

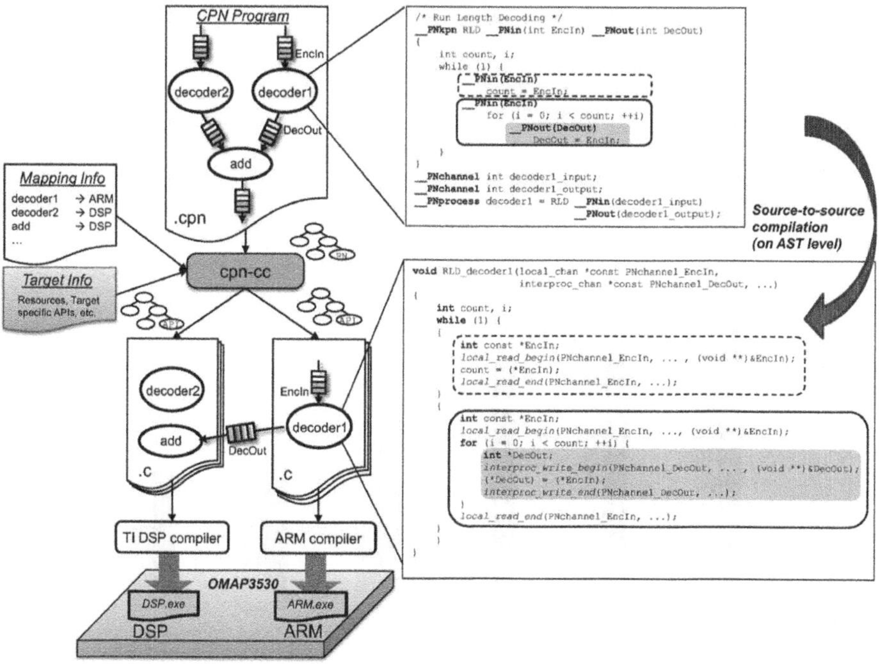

Figure 10.19. *A complete compilation framework for OMAP3530 using MAPS*

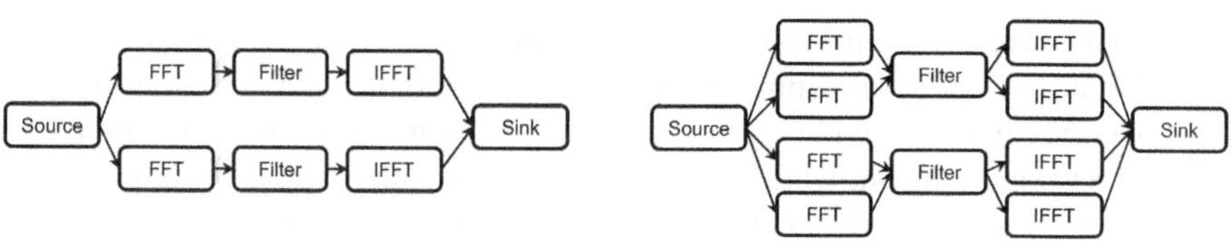

(a) Audio-1 (b) Audio-2

Figure 10.20. *Benchmark: digital audio filter*

The second is a radar application to detect moving objects on the ground from the air, by sending periodic radar pulses. The benchmark was also initially available as a sequential C code. The first conversion into a CPN program took half a day.

Figure 10.22 shows the graphical representation of the first version. The topology of the parallelized version basically follows the algorithmic division of the application. Similarly, we performed the mapping step to the C6678 platform, and the results are shown in Figure 10.22(a). It is clear that the performance of the first process network version of the radar application did not scale well beyond four cores. The speedup stayed about the same from four cores to eight cores, which indicated that the efficiency of the parallel platform usage actually decreased.

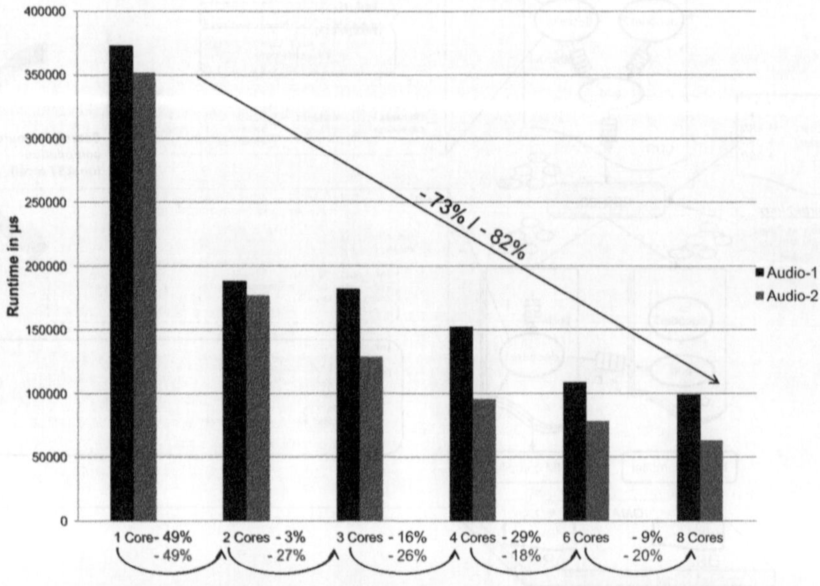

Figure 10.21. *Results of the audio filter (the percentage indicates the reduction of runtime between two mappings)*

We have investigated this saturation and found the reason to be that, the original algorithmic division of the application led to a process network topology that has large deviance in the individual process runtime. This indicates that some processes took a long time to finish, while others needed a relatively short time. In particular, the processes Calculate Filter and Apply Filter (shown in gray in Figure 10.22) are the main number-crunching tasks in this application. This imbalance led to an inefficient usage of the available cores on the C6678 platform.

Similar problems often arise in the process of refining the algorithmic specification for optimized performance, towards a particular parallel architecture. We attempted to solve this by merging the tasks that require a short time to finish and splitting the long-running processes into small ones. The processes with the same boxing style in Figure 10.22 were merged into one process, for example Pulse Compression,

Reordering2, Reordering1 and Calculate Steer Vectors. The processes, Calculate Filter and Apply Filter which have data-level parallelism, are split into four smaller processes. The changes were implemented using the CPN language with very little effort. The improved process network topology is shown in Figure 10.23(a). We then used the same mapping heuristic to generate the spatial mapping for the improved radar application. Figure 10.24(b) showed that performance figures such as the speedup and efficiency are much better than those from the initial version. Figure 10.23(b)–(d) plotted the detailed spatial mapping results for two, four and eight cores, respectively. The process merging and splitting allowed further exploitation of the computing resources of the C6678 platform.

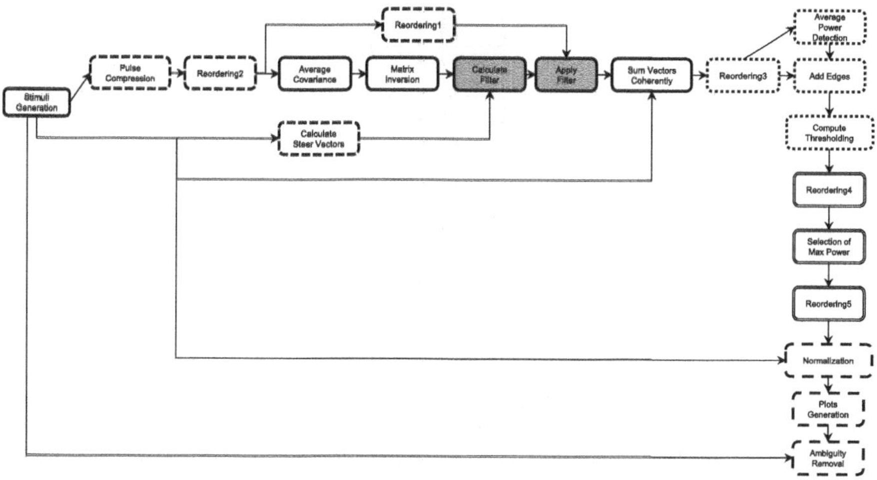

Figure 10.22. *Benchmark: radar application*

The two case studies represented the typical design space exploration process of parallelizing applications for multi-core platforms. The CPN language not only eases the initial conversion from sequential code into parallelized form, but also facilitates optimizing the process networks incrementally. Software developers' productivity is also significantly enhanced as the compilation process is automated. The first case study (digital audio filter) took around one person-day, to finish parallelization and mapping onto the TI platform for each parallelized version. As a comparison, we have also attempted the same procedure for the digital audio filter application manually, using the native tools. It costs a week or so to evaluate just one data point (two cores) in Figure 10.21. The majority of the overhead spent, was on the debugging effort. We have estimated that for the audio filter case, the productivity improvement is around 30x–150x, subject to the programmer's experience level. The more complex the applications are, the more productivity the programmer will gain. Last but not least, programs written in CPN are also portable to other MPSoC platforms.

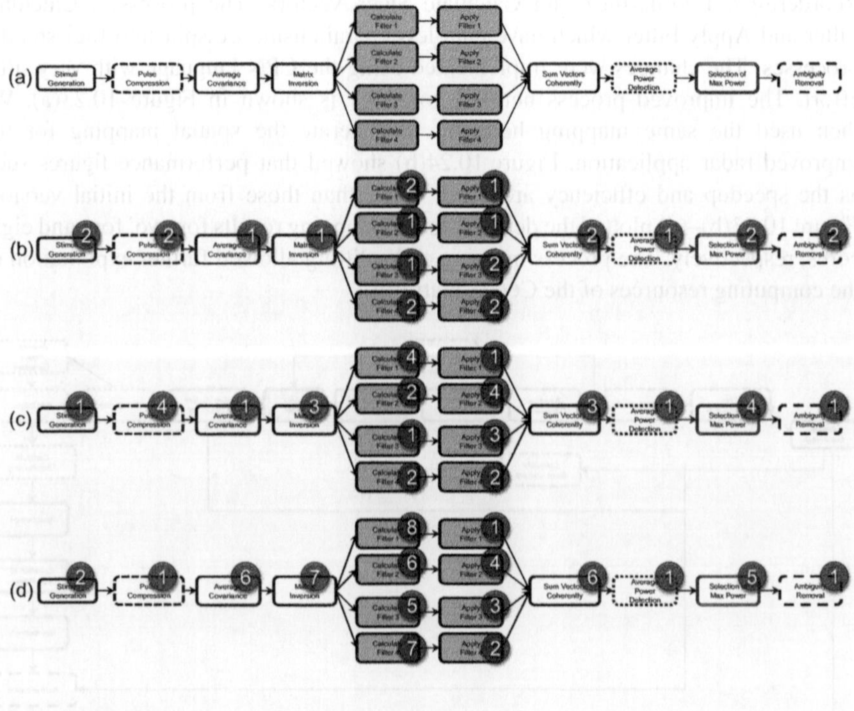

Figure 10.23. *Mapping results of the radar application: improved version. (a) Process network topology (only edges that communicate major data traffic are plotted for the sake of clarity. (b) Mapping onto two cores. (c) Mapping onto four cores. (d) Mapping onto eight cores*

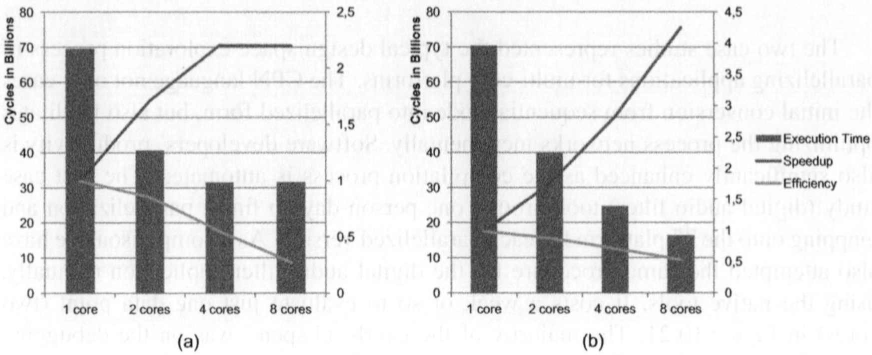

Figure 10.24. *Results of the radar applications: performance. (a) Initial version and (b) improved version.* $Speedup = \frac{Execution\ Time}{Sequential\ Time}$; $Efficiency = \frac{Speedup}{Number\ of\ cores\ used}$

10.5.2. *Retargetability*

One of the main parameters of different MAPS instances is the so-called target info shown on the left side of Figure 10.19. It contains the target-specific information, such as, available resources (processors, inter-communication schemes) and API calls, which needs to be known to the source-to-source compiler. For retargeting MAPS towards different MPSoC platforms, it is essential to update this information. Our experience on the retargetability of the MAPS infrastructure is reported in this section.

Table 10.1 summarizes the retargeting process that was carried out towards the reported MPSoC platforms. The main types of information required for retargeting are the following:

– Platform details: this includes how many programmable processors the target platform has. This reveals where it is possible to simultaneously execute concurrent processes to the compiler.

– Multi-tasking runtime: this indicates how the runtime environments of processors of the target MPSoC support concurrent processes. Often in cases such as that of an OS like Linux or proprietary ones running on a processor, the OS has multi-tasking APIs to manage concurrent tasks (or processes).

– Communication: processes of streaming applications run simultaneously while communicating with one another. Therefore, it is necessary for the compiler to know the possible ways in which communication between processes may be realized. This also includes the type (inter- or intra- processor) and specific API calls.

	TI OMAP3530	TI C6678	Synopsys MCO	Pthreads
Platform details	ARM + DSP	*N*x DSPs	*N*x VPUs	Multi-cores supporting Pthreads
Multi-tasking Runtime	OS thread (ARM and DSP)	OS thread	MCO task modules	OS thread
Communication	shared memory	shared memory or message-passing	shared memory or message-passing	shared-memory
Retargeting effort	20*d*	20*d*	10*d*	5*d*

Table 10.1. *Retargeting MAPS towards MPSoC platforms*

The actual retargeting process consists of using this target-specific information in various locations within the MAPS infrastructure. The effort that we invested in supporting these platforms is reported in Table 10.1. It ranges from 20 person-days to 5 person-days, depending on the complexity of target platforms. The time also includes the initial time required to learn the platform details. This effort is acceptable and justified by the increase in the automation level of the compilation process.

10.6. Conclusion

More than a decade ago, major computing processor manufacturers began to integrate multiple (simple) cores into a single chip, namely multi-cores, to maintain scaling according to Moore's law. While the transition from scalar (uni)processors to multi-cores is something of an evolutionary step in terms of hardware, it has instigated fundamental changes in software development. Things got even worse when requirements are so different from the HPC to embedded systems and more and more heterogeneity has been seen in the hardware. As commonly agreed, the software development process for multi-cores has to be revolutionized. Though many research efforts have been made to provide solutions for developing the multi-core software, lacking practical tooling support has seriously jeopardized maximizing the potential of multi-cores.

This work describes our systematic approach to tackling the multi-core programming challenge from a practical perspective. A compiler infrastructure for heterogeneous embedded multi-cores has been developed. As proof and validation of this work, the tooling infrastructure has been applied successfully in many academic and industrial case studies of multi-core design and development. A clean, lightweight C language extension called CPN (C for Process Networks) is developed to capture streaming models which are common in embedded applications. It is designed to keep the syntax as close to C as possible while making process networks structured and readable. A minimum set of new keywords is added to the C language to describe concurrent processes and channels, which act as the communication between processes. A source-to-source compiler, cpn-cc, was developed as the core component for a multi-core compiler framework. The implementation of cpn-cc is based on the Clang/LLVM by mainly using the AST transformations in the frontend. Unlike the compilers for scalar processors, the cpn-cc does not only need CPN programs as input, but also, a mapping info that specifies the spatial and temporal mapping of processes onto processing elements, available in the target multi-core platform. The cpn-cc and surrounding software components in the framework are made to be extensible and customizable to suit different requirements in the multi-core design practice. Several real-world multi-core platforms, such as TI OMAP 3530 and TI C6678 as well as system-level virtual platforms such as Synopsys MCO, have been used as target platforms successfully. The results of mapping many benchmarks onto multi-core hardware are presented. The applications have achieved good speed-ups and software development productivity has also been greatly improved. In addition, advanced use cases, such as automatic calibration of streaming applications for software mapping exploration and the legacy software migration for a tablet (Sheng *et al.* 2013), have demonstrated the versatility of the multi-core compiler infrastructure developed in this work (Sheng *et al.* 2014). Finally, we discussed the state-of-the-art in software distribution, detailing the mapping flow of the MAPS framework and providing insight into current research directions.

10.7. References

Aho, A.V., Lam, M.S., Sethi, R., and Ullman, J.D. (2006). *Compilers: Principles, Techniques, and Tools*, 2nd edition. Prentice Hall.

Bamakhrama, M.A., Zhai, J.T., Nikolov, H., and Stefanov, T. (2012). A methodology for automated design of hard-real-time embedded streaming systems. *DATE 2012: Proceedings of the 15th Design, Automation, and Test in Europe conference.*

van Berkel, C.H.K. (2009). Multi-core for mobile phones. *Proceedings of the Conference on Design, Automation and Test in Europe, DATE '09, European Design and Automation Association*, Leuven, Belgium, 1260–1265.

Bhattacharyya, S.S., Deprettere, E.F., Leupers, R., and Takala, J. (2010). *Handbook of Signal Processing Systems*. Springer.

Bilsen, G., Engels, M., Lauwereins, R., and Peperstraete, J. (1996). Cyclo-static dataflow. *IEEE Transactions on Signal Processing*, 44(2), 397–408.

Brunet, S.C. (2015). Analysis and optimization of dynamic dataflow programs. PhD Thesis, Ecole Polytechnique Fédérale de Lausanne (EPFL).

Castrillon, J., Velásquez, R., Stulova, A., Sheng, W., Ceng, J., Leupers, R., Ascheid, G., and Meyr, H. (2010). Trace-based KPN composability analysis for mapping simultaneous applications to MPSoC platforms. *Proceedings of the Conference on Design, Automation and Test in Europe, DATE '10*, European Design and Automation Association, Leuven, Belgium, 753–758 [Online]. Available at: http://dl.acm.org/citation.cfm?id=1870926.1871107.

Castrillon, J., Sheng, W., and Leupers, R. (2011). Trends in embedded software synthesis. *Proceedings of the International Conference Embedded Computer Systems: Architecture, Modeling and Simulation (SAMOS), 2011, IEEE*, 347–354.

Castrillon, J., Tretter, A., Leupers, R., and Ascheid, G. (2012). Communication-aware mapping of KPN applications onto heterogeneous MPSoCs. *DAC*.

Castrillon, J., Leupers, R., and Ascheid, G. (2013). MAPS: Mapping concurrent dataflow applications to heterogeneous MPSoCs. *IEEE Transactions on Industrial Informatics*, 9(1), 527–545.

Castrillon Mazo, J. (2013). Programming heterogeneous MPSoCs: Tool flows to close the software productivity gap. PhD Thesis [Online]. Available at: http://publications.rwth-aachen.de/record/211242/files/4534.pdf.

Cisco Systems (2017). Cisco Visual Networking Index: Global Mobile 2017.

Dart, D. (2001). DSP/BIOS Kernel Technical Overview. Texas Instruments Application Report (SPRA780).

Das, A., Singh, A.K., and Kumar, A. (2015). Execution trace–driven energy-reliability optimization for multimedia MPSoCs. *ACM Transactions on Reconfigurable Technology and Systems*, 8(3), 18:1–18:19.

Eker, J. and Janneck, J.W. (2003). CAL language report specification of the CAL actor language. Technical Report UCB/ERL M03/48, EECS Department, University of California, Berkeley [Online]. Available at: http://www.eecs.berkeley.edu/Pubs/TechRpts/2003/4186.html.

Eusse, J.F., Williams, C., and Leupers, R. (2014). CoEx: A novel profiling-based algorithm/architecture Co-exploration for ASIP design. *ACM Transactions on Reconfigurable Technology and Systems*.

Gao, L., Huang, J., Ceng, J., Leupers, R., Ascheid, G., and Meyr, H. (2009). TotalProf: A fast and accurate retargetable source code profiler. *CODES+ISSS '09: Proceedings of the 7th IEEE/ACM international conference on Hardware/software codesign and system synthesis*, ACM, New York, NY, USA, 305–314.

Giannopoulou, G., Poplavko, P., Socci, D., Huang, P., Stoimenov, N., Bourgos, P., Thiele, L., Bozga, M., Bensalem, S., Girbal, S., Faugere, M., Soulat, R., and Dupont de Dinechin, B. (2016). DOL-BIP-Critical: A tool chain for rigorous design and implementation of mixed-criticality multi-core systems. Technical Report 363, ETH Zurich, Laboratory TIK.

Goens, A. and Castrillon, J. (2015). Analysis of process traces for mapping dynamic KPN applications to MPSoCs. In *System Level Design from HW/SW to Memory for Embedded Systems. IESS 2015. IFIP Advances in Information and Communication Technology*, vol 523, Götz, M., Schirner, G., Wehrmeister, M.A., Al Faruque, M.A., Rettberg, A. (eds). Springer International Publishing, Foz do Iguaçu, Brazil, 116–127.

Goens, A., Khasanov, R., Castrillon, J., Polstra, S., and Pimentel, A. (2016). Why comparing system-level MPSoC mapping approaches is difficult: A case study. *Proceedings of the IEEE 10th International Symposium on Embedded Multicore/Many-core Systems-on-Chip (MCSoC-16)*, Ecole Centrale de Lyon, Lyon, France, 281–288.

Goens, A., Khasanov, R., Hähnel, M., Smejkal, T., Härtig, H., and Castrillon, J. (2017a). TETRiS: A multi-application run-time system for predictable execution of static mappings. *Proceedings of the 20th International Workshop on Software and Compilers for Embedded Systems (SCOPES'17), SCOPES '17*, ACM, New York, NY, USA, 11–20 [Online]. Available: http://doi.acm.org/10.1145/3078659.3078663.

Goens, A., Siccha, S., and Castrillon, J. (2017b). Symmetry in software synthesis. *ACM Transactions on Architecture and Code Optimization (TACO)*, 14(2), 20:1–20:26 [Online]. Available at: http://doi.acm.org/10.1145/3095747.

Hascoët, J., Desnos, K., Nezan, J., and Dupont de Dinechin, B. (2017). Hierarchical dataflow model for efficient programming of clustered manycore processors. *2017 IEEE 28th International Conference on Application-specific Systems, Architectures and Processors (ASAP)*, 137–142.

Haubelt, C., Falk, J., Keinert, J., Schlichter, T., Streubühr, M., Deyhle, A., Hadert, A., and Teich, J. (2007). A SystemC-based design methodology for digital signal processing systems. *EURASIP Journal on Embedded Systems*, 2007(1), 22.

International Technology Roadmap for Semiconductors (2011). System drivers. *International Technology Roadmap for Semiconductors 2011 Edition.*

Kahn, G (1974). The semantics of a simple language for parallel programming. *IFIP Congress 74*, North Holland, Amsterdam, 471–475.

Keinert, J., Streubühr, M., Schlichter, T., Falk, J., Gladigau, J., Haubelt, C., Teich, J., and Meredith, M. (2009). SystemCoDesigner – An automatic ESL synthesis approach by design space exploration and behavioral synthesis for streaming applications. *ACM Transactions on Design Automation of Electronic Systems*, 14, 1:1–1:23 [Online]. Available at: http://doi.acm.org/10.1145/1455229.1455230.

Kempf, T., Doerper, M., Leupers, R., Ascheid, G., Meyr, H., Kogel, T., and Vanthournout, B. (2005). A modular simulation framework for spatial and temporal task mapping onto multi-processor SoC platforms. *Proceedings of the conference on Design, Automation and Test in Europe*, 876–881.

Khasanov, R. and Castrillon, J. (2020). Energy-efficient runtime resource management for adaptable multi-application mapping. *Proceedings of the 2020 Design, Automation and Test in Europe Conference (DATE), DATE '20*, EDA Consortium.

Khasanov, R., Goens, A., and Castrillon, J. (2018). Implicit data-parallelism in Kahn process networks: Bridging the MacQueen gap. *Proceedings of the 9th Workshop and 7th Workshop on Parallel Programming and RunTime Management Techniques for Manycore Architectures and Design Tools and Architectures for Multicore Embedded Computing Platforms (PARMA-DITAM'18), co-located with 13th International Conference on High-Performance and Embedded Architectures and Compilers (HiPEAC), PARMA-DITAM '18*, ACM, New York, NY, USA, pp. 20–25 [Online]. Available at: http://doi.acm.org/10.1145/3183767.3183790.

Kloss, N. (2003). Application programming strategies for TI's OMAP solutions. *Embedded Edge.*

Lattner, C. (2008). LLVM and Clang: Next generation compiler technology. *The BSD Conference*, Ottawa, Canada.

Lee, E.A. and Messerschmitt, D.G. (1987). Synchronous data flow. *Proceeding of the IEEE*, vol. 75.

Leupers, R. and Castrillon, J. (2010). MPSoC programming using the MAPS compiler. *Design Automation Conference (ASP-DAC), 2010 15th Asia and South Pacific*, 897–902.

Leupers, R., Aguilar, M.A., Eusse, J.F., Castrillon, J., and Sheng, W. (2017). *MAPS: A Software Development Environment for Embedded Multicore Applications.* Springer Netherlands, 1–33.

Marwedel, P., Bacivarov, I., Lee, C., Teich, J., Thiele, L., Xu, Q., Kouveli, G., Ha, S., and Huang, L. (2011). Mapping of applications to MPSoCs. *2011 Proceedings of the 9th International Conference on Hardware/Software Codesign and System Synthesis (CODES+ISSS)*, 109–118.

Nikolov, H. (2009). System-Level design methodology for streaming multi-processor embedded systems. PhD Thesis, Universiteit Leiden.

Odendahl, M., Castrillon, J., Volevach, V., Leupers, R., and Ascheid, G. (2013). Split-cost communication model for improved MPSoC application mapping. *Proceedings of the International Symposium on System on Chip (SoC)*, 2013, IEEE, 1–8.

Page, M., Rossi, L., and Rand, C. (2010). A viable future model for the Internet. Technical Report, A. T. Kearney Paper.

Pelcat, M., Desnos, K., Heulot, J., Guy, C., Nezan, J.-F., and Aridhi, S. (2014). PREESM: A dataflow-based rapid prototyping framework for simplifying multicore DSP programming. *2014 6th European Embedded Design in Education and Research Conference (EDERC)*, IEEE, 36–40.

Pimentel, A.D., Erbas, C., and Polstra, S. (2006). A systematic approach to exploring embedded system architectures at multiple abstraction levels. *IEEE Transactions on Computers*, 55(2), 99–112.

Ptolemaeus, C., (ed.) (2014). *System Design, Modeling, and Simulation using Ptolemy II*, Ptolemy.org [Online]. Available at: http://ptolemy.org/books/Systems.

Quan, W. and Pimentel, A.D. (2013). A scenario-based run-time task mapping algorithm for MPSoCs. *Proceedings of the 50th Annual Design Automation Conference, DAC '13*, ACM, New York, NY, USA, pp. 131:1–131:6 [Online]. Available at: http://doi.acm.org/10.1145/2463209.2488895.

Quan, W. and Pimentel, A.D. (2014). Towards exploring vast MPSoC mapping design spaces using a Bias-elitist evolutionary approach. *DSD 2014*.

Quan, W. and Pimentel, A.D. (2016). Scenario-based run-time adaptive MPSoC systems. *Journal of Systems Architecture*, 62, 12–23.

Schirrmeister, F. and Sheridan, P. (2011). Optimizing multicore system performance. *Synopsys Insight, Issue 1*.

Schor, L., Bacivarov, I., Rai, D., Yang, H., Kang, S.-H., and Thiele, L. (2012). Scenario-based design flow for mapping streaming applications onto on-chip many-core systems. *Proceedings of the 2012 International Conference on Compilers, Architectures and Synthesis for Embedded Systems, CASES '12*, ACM, New York, NY, USA, 71–80 [Online]. Available at: http://doi.acm.org/10.1145/2380403.2380422.

Schwarzer, T., Roloff, S., Richthammer, V., Khaldi, R., Wildermann, S., Glass, M., and Teich, J. (2018). On the complexity of mapping feasibility in many-core architectures. *2018 IEEE 12th International Symposium on Embedded Multicore/Many-core Systems-on-Chip (MCSoC)*, 176–183.

Sheng, W., Szymanski, P., Leupers, R., and Ascheid, G. (2013). Software migration for parallel execution on a multicore tablet: A case study. *IEEE 7th International Symposium on Embedded Multicore SoCs (MCSoC-13)*.

Sheng, W., Schürmans, S., Odendahl, M., Bertsch, M., Volevach, V., Leupers, R., and Ascheid, G. (2014). A compiler infrastructure for embedded heterogeneous mpsocs. *Parallel Computing*, vol. 40, Elsevier, 51–68.

Sriram, S. and Bhattacharyya, S.S. (2009). *Embedded Multiprocessors: Scheduling and Synchronization*, 2 edition. Marcel Dekker, Inc., New York, NY, USA.

Stuijk, S., Geilen, M., Theelen, B., and Basten, T. (2011). Scenario-aware dataflow: Modeling, analysis and implementation of dynamic applications. *2011 International Conference on Embedded Computer Systems (SAMOS)*, 404–411.

Texas Instruments (2011). C6678 multicore fixed and floating-point digital signal processor [Online]. Available at: http://ti.com/product/tms320c6678 #technicaldocuments/ (accessed 06/2012).

Thiele, L., Bacivarov, I., Haid, W., and Huang, K. (2007). Mapping applications to tiled multiprocessor embedded systems. *ACSD '07: Proceedings of the Seventh International Conference on Application of Concurrency to System Design*, IEEE Computer Society, Washington, DC, USA, 29–40.

Weichslgartner, A., Wildermann, S., Götzfried, J., Freiling, F., Glaß, M., and Teich, J. (2016). Design-time/Run-time mapping of security-critical applications in heterogeneous MPSoCs. *Proceedings of the 19th International Workshop on Software and Compilers for Embedded Systems, SCOPES '16*, ACM, New York, NY, USA, pp. 153–162 [Online]. Available at: http://doi.acm.org/10.1145/2906363.2906370.

Wiggers, M.H. (2009). Aperiodic multiprocessor scheduling for real-time stream processing applications. PhD Thesis, University of Twente. 978-90-365-2850-4.

List of Authors

Liliana ANDRADE
TIMA Lab
Université Grenoble Alpes
France

Caaliph ANDRIAMISAINA
LIST
CEA
Gif-sur-Yvette
France

Alexandre CARBON
LIST
CEA
Gif-sur-Yvette
France

Jeronimo CASTRILLON
Chair for Compiler Construction
TU Dresden
Germany

Amir CHARIF
LIST
CEA
Gif-sur-Yvette
France

Julian CHESTERFIELD
Sunlight.io Limited
Cambridge
United Kingdom

Marcello COPPOLA
STMicroelectronics
Grenoble
France

Stefan A. DAMJANCEVIC
Vodafone Chair Mobile
Communications Systems
TU Dresden
Germany

Raphaël DAVID
LIST
CEA
Gif-sur-Yvette
France

Philippe DORE
LIST
CEA
Gif-sur-Yvette
France

Paul DUBRULLE
LIST
CEA
Gif-sur-Yvette
France

Babak FALSAFI
School of Computer and Communication Sciences
EPFL
Lausanne
Switzerland

Gerhard P. FETTWEIS
Vodafone Chair Mobile Communications Systems
and
Barkhausen Institut
TU Dresden
Germany

Michail FLOURIS
Sunlight.io Greece
Heraklion
Greece

Etienne HAMELIN
LIST
CEA
Gif-sur-Yvette
France

Matthias HERRMANN
Microelectronic Systems Design Research Group
TU Kaiserslautern
Germany

Ahmed JERRAYA
Cyber Physical Systems Programs
CEATech
Grenoble
France

George KORNAROS
ISCA Lab
Hellenic Mediterranean University
Heraklion
Greece

Laurent LE GARFF
System Engineering Division
Renault
Guyancourt
France

Rainer LEUPERS
Institute for Communication Technologies and Embedded Systems
RWTH Aachen University
Germany

Shuai LI
LIST
CEA
Gif-sur-Yvette
France

Emil MATUS
Vodafone Chair Mobile Communications Systems
TU Dresden
Germany

Maroun OJAIL
LIST
CEA
Gif-sur-Yvette
France

Alexis OLIVEREAU
LIST
CEA
Gif-sur-Yvette
France

Frédéric ROUSSEAU
TIMA Lab
Université Grenoble Alpes
France

Salah-Eddine SAIDI
LIST
CEA
Gif-sur-Yvette
France

Tanguy SASSOLAS
LIST
CEA
Gif-sur-Yvette
France

Weihua SHENG
Hong Kong Research Center
Huawei
Hong Kong
China

Dmitry UTYANSKY
Synopsys
Saint Petersburg
Russia

Kees VAN BERKEL
Department of Mathematics and
Computer Science
Eindhoven University of
Technology
The Netherlands

Pieter VAN DER WOLF
Solutions Group
Synopsys, Inc.
Eindhoven
The Netherlands

Nicolas VENTROUX
LIST
CEA
Gif-sur-Yvette
France

Norbert WEHN
Microelectronic Systems Design
Research Group
TU Kaiserslautern
Germany

Arief WICAKSANA
TIMA Lab
Université Grenoble Alpes
France

Marilyn WOLF
Department of Computer Science
and Engineering
University of Nebraska–Lincoln
USA

Author Biographies

Caaliph Andriamisaina received an MSc degree in Electronics and Telecommunications Technology from the University of Western Brittany in Brest, France, in 2004, and a PhD degree in Computer Science and Electronic Engineering from the University of South Brittany in Lorient, France, in 2008. From 2008 to 2010, he was a post-doctoral researcher with the Lab-STICC, University of South Brittany, working on high-level synthesis algorithms development. In 2010, he joined CEA List Institute, Saclay, France, and since then, he has been working on the development of embedded multi-core architectures. His current research interests include methodologies for power analysis optimization, co-emulation and hardware accelerators.

Alexandre Carbon was a CEA Expert and a lead system architect in charge of research development projects in electronics. He received a PhD degree in Computer Science from University Pierre et Marie Curie (Paris VI) in 2013. He joined the CEA List as a research engineering specializing in artificial intelligence. He was especially involved in the development of a low-power embedded processor for deep learning applications (PNeuro) and the management of its integration (ASIC FPGA). He was also a referent for neural hardware architecture for the Architecture, IC Design Embedded Software Division, and in charge of competitor solution analysis. In 2017, he was the hardware architecture work package Leader in the Future architecture for Automotive Computer Environment (FACE) project with Renault, where he managed a 15-people multi-disciplinary team covering activities from board conception to low-level software integration through architecture definition and FPGA system development. He was also in charge of the hardware system architecture definition and validation with Renault experts and in charge of the relationships with industrial partners and suppliers. He participated in many congresses and exhibitions (e.g. CES'19), has authored or co-authored 11 international publications and held seven patents. He is now an Electronic Project Manager in the aeronautics industry.

Jeronimo Castrillon is a professor in the Department of Computer Science at the TU Dresden, where he is also affiliated with the Center for Advancing Electronics Dresden (CfAED). He received an Electronics Engineering degree from the Pontificia Bolivariana University, Colombia, in 2004, a master's degree from

Multi-Processor System-on-Chip 2 – Applications,
coordinated by Liliana Andrade and Frédéric Rousseau. © ISTE Ltd 2020.

the ALaRI Institute, Switzerland, in 2006 and a PhD degree (Dr.-Ing.) with honors from the RWTH Aachen University, Germany, in 2013. His research interests lie on methodologies, languages, tools and algorithms for programming complex computing systems. He has more than 80 international publications and has been a member of technical program and organization committees in international conferences and workshops (e.g. DAC, DATE, ESWEEK, CGO, LCTES, Computing Frontiers, ICCS and FPL). He is also a regular reviewer for ACM and IEEE journals (e.g. IEEE TCAD, IEEE TPDS, ACM TODAES and ACM TECS). In 2014, Professor Castrillon co-founded Silexica GmbH/Inc., a company that provides programming tools for embedded multi-core architectures. He is Senior Member IEEE, Member ACM and was a founding member (2017–2019) of the executive committee of the ACM "Future of Computing Academy" (FCA).

Amir CHARIF obtained his master's degree in Computer Engineering and Electronics from the Polytechnical School of the University of Grenoble (Polytech'Grenoble) in 2014. Then, he pursued his PhD at the TIMA Laboratory in Grenoble, where he made several contributions to the field of networks-on-chip. He obtained his PhD in 2017 and joined CEA List the same year to work on simulation and virtual prototyping. His research interests include computer architecture, high-performance simulation and design automation.

Julian CHESTERFIELD is the CEO of Sunlight.io Ltd. He studied Computer Science at University College London (MSc) and went on to complete a PhD at the University of Cambridge. He was an early member of XenSource, a Cambridge University spin-off company that developed Xen, the open-source hypervisor acquired by Citrix in 2008. He was the storage development team lead and senior architect for the XenServer and OSS Xen products. In 2011, he joined OnApp Ltd as the Chief Scientific Officer, responsible for the R&D activities, as well as the development of the Enterprise Cloud Platform. Since 2018, he has been the Founder and CEO of Sunlight.io.

Marcello COPPOLA is a Technical Director at STMicroelectronics and has more than 20 years of industry experience focused on developing breakthrough technologies and knowledge in EU-related matters concerning R&D funding. He has a graduate degree in Computer Science from the University of Pisa, Italy. His research interests include AI, IoT for education, cyber–physical systems, 5G, automotive technologies and multi-core and many-core SoCs. Coppola has co-authored more than 50 scientific publications and held various roles in top international conferences and workshops. He holds 26 patents and is involved in multiple European research projects.

Stefan A. DAMJANCEVIC is a junior scientist at TU Dresden Vodafone Chair Mobile Communications Systems within the signal processing hardware group. He is currently pursuing his PhD degree under the supervision of Professor Gerhard P. Fettweis. His research is on the implementation aspects of 5G and beyond physical layer algorithms and modem architectures. He earned his electrical engineering Dipl.-Ing. degree from the University of Novi Sad and an MSc from TU Dresden. He is an alumni of the German Academic Exchange Service (DAAD) and the Gustav Stresemann Institute (GSI).

Raphaël DAVID is a CEA Senior Expert in charge of strategic and transversal programs for the CEA List. He received his PhD degree in Computer Science from the University of Rennes I in 2003. He joined the CEA List to design programmable and reconfigurable multi- or many-core architectures to support variable execution conditions, either coming from technology or data-dependent applications. He has authored or co-authored more than 50 international publications and holds eight patents in the field. In 2010, he took the lead of the Embedded Computing Laboratory and managed a large research team devoted to the design of embedded vision computers. In 2015, he built a new lab to benefit from the cross-fertilization of HW and embedded SW skills as well as to promote efficient mix-critical embedded computers and associated programming and design frameworks. He has been attached to the Director of the CEA List since 2017 to build and manage a multi-disciplinary team, spread across eight laboratories and three divisions, and devoted to the design of the Future architecture for Automotive Computer Environment (FACE) project with Renault.

Philippe DORE graduated from telecom ParisTech and is a researcher at the Commissariat de l'Energie Atomique, the French DoE, where he currently manages projects around neural networks IA systems. Prior to his current position, after spending 18 years in the telecom industry, mainly as a system architect, he contributed to the development of a complete industry-grade dataflow compiler for the 256 cores MPPA architecture as part of a joint CEAKALRAY lab. Then, for about one year, he led a research team responsible for developing an industry-grade tool suite used to design safety-critical applications for various multi-cores. Since 2013, he has managed several research projects and elaborated a Model of Computation and Communication for real-time, speculative dataflow applications with deterministic guarantees. In 2017, he took the lead of the "Prototyping and Validation" activities of the Future architecture for Automotive Computer Environment (FACE) project with Renault.

Paul DUBRULLE is a CEA Expert in Formal Methods and Software Engineering, working as a research engineer at the Architecture, IC Design & Embedded Software Division. He has a master's degree in Computer Science from the University of Maine, France, and started his career at CEA List in 2008. Since then, he has been working on languages, models, compilers and runtimes for mixed-critical systems, developing integrated solutions to enhance the reliability and performance of embedded software. He assumed the responsibility of work package and task leader in both collaborative and industrial projects, including the development of an embedded runtime and code generation tools in a dataflow compiler for the Kalray MPPA Chip and safe hypervision services for mixed-criticality in the ARTEMIS

EMC^2 project. In 2017, he joined the Future architecture for Automotive Computer Environment (FACE) project with Renault, as a cross-disciplinary expert. He defined the Polygraph Model of Computation and envisioned the integration model based on behavioral contracts for software components, seamlessly compatible with the automotive standard AUTOSAR. He worked with all the technical work packages to ensure coherence in the definition of the end-to-end software design methodology and in the integration of the tools and the embedded platform. He authored 14 international publications and four patents.

Babak FALSAFI is Professor at the School of Computer and Communication Sciences and the Founding Director of the EcoCloud research center at EPFL. He has worked on server architecture for quite some time and has made contributions to a few industrial platforms including the WildFire/WildCat family of multi-processors by Sun Microsystems (now Oracle); memory system technologies for IBM BlueGene/P and Q and ARM cores; and server evaluation methodologies in use by AMD, HP and Google (PerfKit). His recent work on scale-out server processor design lays the foundation for Cavium ThunderX1. He is a fellow of ACM and IEEE.

Gerhard P. FETTWEIS has been the Vodafone Chair Professor at TU Dresden since 1994 and has been Head of the Barkhausen Institute since 2018. He earned his PhD under H. Meyr's supervision from RWTH Aachen in 1990. After one year at IBM Research in San Jose, CA, he moved to TCSI, Inc., Berkeley, CA. He coordinates the 5G Lab Germany and has coordinated two German Science Foundation (DFG) centers at TU Dresden, namely cfaed and HAEC. In 2019, he was elected into the DFG Senate. His research focuses on wireless transmission and chip design for wireless/IoT platforms, with 20 companies from Asia/Europe/the United States sponsoring his research. He also serves on the board of National Instruments Corp and advises other companies. Gerhard is IEEE Fellow, member of the German Academy of Sciences (Leopoldina), the German Academy of Engineering (acatech) and received multiple IEEE recognitions and also received the VDE ring of honor. In Dresden, his team has spun out 17 start-ups and set up funded projects in volume of close to EUR 1/2 billion. He co-chairs the IEEE 5G/Future Networks Initiative, and has helped organize IEEE conferences, most notably as TPC Chair of ICC 2009 and of TTM 2012, and as General Chair of VTC Spring 2013 and DATE 2014.

Michail FLOURIS is the VP of Product Architecture at Sunlight.io. He received his PhD in Computer Science from the University of Toronto in 2009 and has over 15 years of experience in high-performance systems engineering. He has over 25 peer-reviewed publications and has been a reviewer for major ACM and IEEE conferences and journals. In 2012, he joined OnApp as a Lead Engineer and Researcher, working on the core components of the Enterprise Cloud Platform. Since 2018, Flouris has joined Sunlight.io and he is currently the VP of Product Architecture, responsible for product architecture, R&D projects and innovation activities.

Etienne HAMELIN graduated from Ecole Polytechnique and Telecom Paristech in 2008. He is a member of the Corps des Mines. Hamelin served for three years as a project officer at the French Ministry of Industry, where he contributed to the development of the French embedded software and systems ecosystem, by supporting numerous collaborative projects from National Competitiveness Clusters, European Clusters and European Joint Undertakings. He then joined CEA List in 2011, where he contributes to both project management and research engineering activities, in various projects ranging from model-based code generation for deterministic real-time applications, to real-time operating systems and middleware development for distributed and multi-core ECUs. Within the Future architecture for Automotive Computer Environment (FACE) project with the Renault–Nissan–Mitsubishi Alliance, Hamelin was the principal software architect and leader of a software development team.

Matthias HERRMANN received his diploma in Communication Engineering from the TU Kaiserslautern. Currently, he is working towards his PhD in Microelectronic System Design in the Department of Electrical Engineering and Information Technology at the University of Kaiserslautern. His special research interests include high-throughput baseband processing.

George KORNAROS is an Associate Professor at the Electrical and Computer Engineering Department, Hellenic Mediterranean University, Heraklion, Greece, where he leads the Intelligent Systems and Computer Architecture Group. In the past, as a hardware designer and system architect, he has designed single-chip network processors for ISD S.A., Ellemedia Tech. Ltd and ICS-FORTH. Currently, he is involved in multiple European research projects. His research interests include multi-/many-core architectures, high-speed communication architectures and energy-efficient and heterogeneous computing. His current research lies on secure and trustworthy computing in IoT. Kornaros has published more than 60 scientific articles and edited a book entitled *MultiCore Embedded Systems*. He holds three patents and is a member of the Technical Chamber of Greece.

Laurent LE GARFF is Innovation Project Manager for alectrical and electronical related projects at Renault–Nissan–Mitsubishi Alliance. He graduated from The Ecole Centrale de Nantes in 2000. Until 2009, he was in charge of high-volume Diesel engine management system and software development and validation. From 2009, he was Gasoline Engine Project Manager in charge of the development and industrialization of a new version of 2.0 liter Renault engine (evolution of mechanic, tuning, regulation and production) dedicated to overseas markets. Since 2015, he has been managing all innovation projects for Alliance related to alectronical architecture, including the Future Architecture for Computing Environment (FACE) project in collaboration with CEA List.

Rainer LEUPERS received MSc (Dipl.-Inform.) and PhD (Dr. rer. nat.) degrees in Computer Science with honors from TU Dortmund in 1992 and 1997. From 1997 to

2001, he was the Chief Engineer at the Embedded Systems Chair at TU Dortmund. In 2002, he joined RWTH Aachen University as a Professor of Software for Systems on Silicon. His research comprises embedded software development tools, multi-core processor architectures, hardware security and system-level electronic design automation. He served in committees of the leading international EDA conferences and received various scientific awards, including Best Paper Awards at DAC and twice at DATE, as well as several industrial awards. Dr. Leupers is also an entrepreneur and actively turns research results into innovations. He holds several patents and is a Co-founder of LISATek (now with Synopsys), Silexica and Secure Elements. As the coordinator of the TETRACOM and TETRAMAX projects, he contributes to EU-wide academia-to-industry technology transfer.

Shuai LI is a researcher at the Embedded and Autonomous Systems Design Laboratory of the CEA List institute in France. He obtained a PhD in Computer Science, specializing on the topic of schedulability analysis of TDMA software radio protocols, while working at THALES and Lab-STICC. His current research is related to model-based software engineering for cyber–physical systems with embedded AI, more specifically real-time software architecture optimization and design and deployment of smart behaviors. He is involved in several industrial and European collaborative R&D projects, where he works on the development of CPS digital platforms, security/performance co-engineering and model-based federation of neural networks in the cloud. He also supervises several PhD students working on the connection between machine learning and software engineering. He is a committer of the open-source Eclipse Papyrus project. Papyrus is a UML modeler and MDE framework for the academic and industrial community. In his free time, he is also a sessional lecturer at the French National Institute of Nuclear Sciences and Techniques (INSTN), part of the Paris-Saclay University. He teaches model-based system engineering with UML, SysML and MARTE.

Emil MATUS has been a senior scientist at Vodafone Chair for Mobile Communications Systems at TU Dresden since 2003, where he is leading the signal processing hardware research group. He received MSc and PhD degrees in EE at TU Kosice (Slovakia), where he focused on wavelet transform and image compression research. His current research interests include wireless physical layer algorithms and communications signal processing hardware architectures. His group introduced several generations of the signal-processing MPSoC architecture called "Tomahawk", which received the DAC/ISSCC Student Design Award 2009, the CoolSilicon Cool Award 2014 and the HIPEAC Award 2017.

Maroun OJAIL is a system architect at CEA List for cyber–physical systems. He received his PhD degree in Computer Science from the University of Rennes I in 2009 for the design of a heterogeneous embedded computing system for multi-standard and multimode mobile terminals. From 2009 to 2012, he was a key developer of the SW framework of a multi-core computing system based on STMicroelectronics STxP70 microcontrollers. From 2013 to 2015, he was the main electronics architect of a state-of-the-art adaptive optical system for the University of Dayton in Ohio, USA. Since 2017, he has been the system architect for the future architecture of the automotive computing environment for the Renault–Nissan–Mitsubishi Alliance.

Alexis OLIVEREAU, CEA Senior Expert in Communication Security and the Internet of Things, is leading the CEA List Communicating Systems Lab. He graduated from the École Nationale Supérieure de l'Électronique et ses Applications, Cergy, France, in 2000. Between 2000 and 2008, he was a research engineer for the Motorola Labs research center in Paris and earned himself a Motorola "GoldBadge" distinction for his patents filing. He joined the Communicating Systems Lab (LSC) of CEA List in January 2009 as a researcher, working on security and privacy aspects of communications in the IoT, industrial and vehicular networks. He coordinated two EU projects dealing with security and resilience of communication services for critical infrastructure protection. In 2017, he was the communication work package Leader in the Future architecture for Automotive Computer Environment (FACE) project with Renault. He has authored more than 30 international publications and held more than 20 patents.

Salah-Eddine SAIDI received master's and PhD degrees in Computer Engineering from Sorbonne University in 2014 and 2018, respectively. As part of the PhD program, he was a research scholar at IFP Energies Nouvelles and Inria from 2014 to 2018, where he contributed to the development of a tool for parallel simulation of cyber–physical systems. Since 2018, he has been with the Embedded Computing Laboratory, CEA Tech, as a postdoctoral researcher. His research interests lie in parallel and multi-core computing, embedded and real-time systems and virtual prototyping.

Tanguy SASSOLAS is a research engineer in the Computing and Design Automation Laboratory at CEA List. He received an MSc and an MEng in Electronics from ENSEEIHT, Toulouse, France, in 2008. His work focused on the design of many-core architectures. In particular, he was deeply involved in the conception of the SESAM multi-processor architecture simulation and exploration environment. His architectural works mainly dealt with the design of hardware support for efficient performance and power management. He then addressed the deficient thermal management design thanks to joint functional/thermal co-simulation. To investigate thermal phenomena timescale, he moved from virtual prototyping to emulation platform and designed efficient co-emulation links for power and thermal emulation. With the same goal, he studied the parallelization of VP simulations and actively participated in the definition of the SCale parallel SystemC kernel. More recently, he also addressed the issue of generic co-simulation links for cyber–physical systems design through the FMI standard.

Weihua SHENG received a bachelor's degree from Tsinghua University in China as well as MSc and PhD (Dr.-Ing.) degrees from RWTH Aachen University in

Germany. From 2003 to 2006, he was a senior R&D engineer at Synopsys. From 2007 to 2013, he spearheaded technology development on multi-core programming tools at the ICE (Institute for Communication Technologies and Embedded Systems) at RWTH Aachen University, which later turned into the foundation of Silexica. In 2014, he co-founded Silexica in Germany. He served as CTO of Silexica between 2014–2016. He worked as GM of APAC in Japan from 2017 to 2019, responsible for all aspects of Silexica sales and operations across the APAC region. He joined Huawei Technologies in September 2019 as Chief Expert in Software Tools and Engineering.

Dmitry UTYANSKY is a senior staff software engineer at SYNOPSYS Corp. (Saint Petersburg, Russia), where he is engaged in digital signal processing algorithms exploration and design and benchmarking of SYNOPSYS DSP architectures. He received an MSc degree in Computer Science at Saint Petersburg State Electrotechnical University "LETI" in 1994 and was subsequently involved in various research and development projects in DSP, audio and image processing and communications. His current interests include algorithms for linear algebra and its applications in RADARs and AI, utilizing vector processors.

Kees VAN BERKEL started his R&D career at Philips Research in 1980, after receiving an MSc degree in EE from TU Delft. From 2000 to mid-2015, he was a fellow at Philips, NXP, ST-Ericsson and Ericsson. He obtained a PhD in CS from TU Eindhoven in 1992, where he has been a Part-time Full Professor since 1996. He pioneered asynchronous VLSI from theory to mass production, as well as embedded vector processing for software-defined radio. His research includes software-defined radio, vector processors, multi-core architectures, resource management and low-power computing. His current research interest is exascale dataflow computing for radio astronomy.

Pieter VAN DER WOLF is a principal product architect at Synopsys. He received his MSc and PhD degrees in Electrical Engineering from Delft University of Technology where he was an Associate Professor before joining Philips Research in 1996. In 2006, he joined NXP Semiconductors when it was spun out of Philips Electronics. In 2009, he joined VirageLogic, which was subsequently acquired by Synopsys. He has worked on a broad range of topics including (multi-)processor architectures and system design methodologies, with a focus on DSP applications.

Nicolas VENTROUX is the Head of the Computing and Design Environment Laboratory at CEA List. He received an MSc and an MEng in Computer Sciences from INSA, Rennes, in 2003, and a PhD degree in Electronics from the University of Rennes in 2006. He is a CEA Senior Expert in multi-processor modeling and design, and was a project manager of many national and European projects. He was a visiting scholar at Carnegie Mellon University (CMU) in the CALCM laboratory in 2012. He started working on SESAM in 2006 and was responsible for its development for almost 10 years. Specifically, he led the design of SCale, a parallel SystemC kernel designed for the acceleration of virtual prototyping. He holds 10 patents and has authored and co-authored several papers in conferences and journals in the multi- and many-core domains. He has also been a reviewer for several international conferences and journals since 2006.

Norbert WEHN holds the Chair for Microelectronic System Design in the Department of Electrical Engineering and Information Technology at the University of Kaiserslautern. He has more than 300 publications in various fields of microelectronic system design and holds several patents. Two start-ups have been spun out of his research group. His special research interests are VLSI-architectures for mobile communication, forward error correction techniques, low-power techniques, advanced SoC and memory architectures, 3D integration, reliability issues in SoC, IoT and hardware accelerators for big data applications.

Arief WICAKSANA obtained his Bachelor degree in Electrical Engineering from the Bandung Institute of Technology (ITB), Indonesia. He later received his Engineering degree and PhD degree in Computer Science from Grenoble Alpes University, France, in 2015 and 2018, respectively. He pursued his PhD degree in the TIMA Laboratory, Grenoble, from 2015 to 2018. Since then, he has been working as a research engineer at Embedded Computing Laboratory, CEA Tech. His research areas include reconfigurable computing, FPGA virtualization and virtual prototyping.

Marilyn WOLF is Professor and Chair of the Department of Computer Science and Engineering at the University of Nebraska–Lincoln. She received her BSc, MSc and PhD in Electrical Engineering from Stanford University in 1980, 1981 and 1984, respectively. She was with AT&T Bell Laboratories from 1984 to 1989. She was on the faculty of Princeton University from 1989 to 2007 and was Farmer Distinguished Chair at Georgia Tech from 2007 to 2019. Her research interests include embedded computing, embedded video and computer vision and VLSI systems. She has received the IEEE Computer Society Goode Memorial Award, the ASEE Terman Award and the IEEE Circuits and Systems Society Education Award. She is a Fellow of the IEEE and ACM and an IEEE Computer Society Golden Core member.

Index